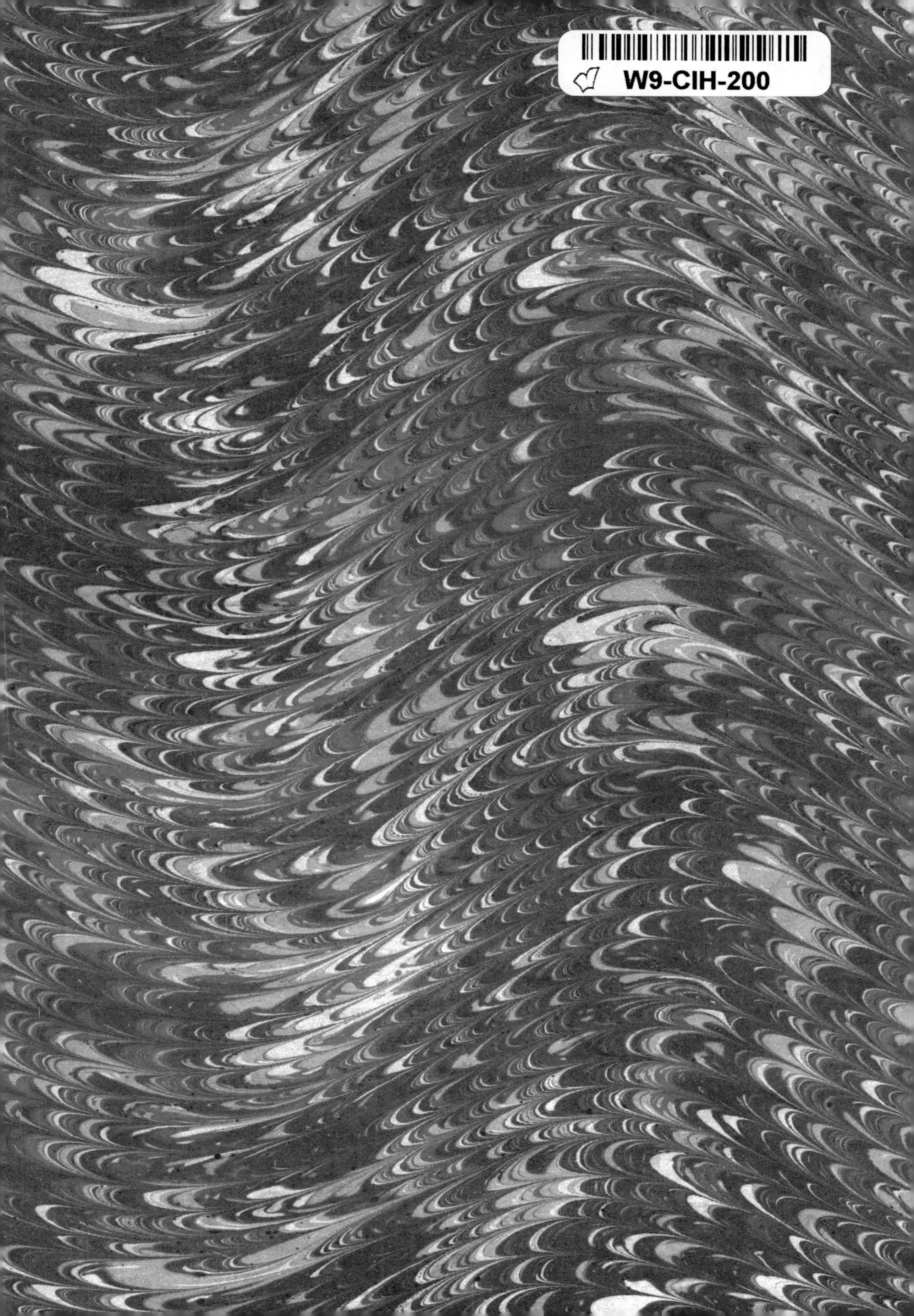
W9-CIH-200

Encyclopedia of Creation Myths

Encyclopedia of Creation Myths

David Adams Leeming
with
Margaret Adams Leeming

ABC-CLIO
Santa Barbara, California
Denver, Colorado
Oxford, England

Library of Congress Cataloging-in-Publication Data

Leeming, David Adams, 1937–
Encyclopedia of creation myths / David Adams Leeming with Margaret Adams Leeming.
Includes bibliographical references and index.
1. Creation—Comparative studies—Encyclopedias. I. Leeming, Margaret Adams, 1968– . II. Title.
BL325.C7L44 1994 291.2'4'03—dc20 94-7169

ISBN 0-87436-739-5

00 99 98 97 96 95 94 10 9 8 7 6 5 4 3 2 1

ABC-CLIO, Inc.
130 Cremona Drive, P.O. Box 1911
Santa Barbara, California 93116-1911

This book is printed on acid-free paper ♾ .
Manufactured in the United States of America

Contents

Introduction

A myth is a narrative projection of a given cultural group's sense of its sacred past and its significant relationship with the deeper powers of the surrounding world and universe. A myth is a projection of an aspect of a culture's soul. In its complex but revealing symbolism, a myth is to a culture what a dream is to an individual.

A creation myth is a cosmogony, a narrative that describes the original ordering of the universe. The word cosmogony is derived from the Greek words kosmos, meaning order, and genesis, meaning birth. A given culture's cosmogony or creation myth describes its sense of how cosmos (order, existence) was established. Just as individuals and families are preoccupied with their origins, cultures need to know where they and the world they live in originated. So it is that virtually all cultures have creation myths.

Like all myths, creation myths are etiological—they use symbolic narrative to explain beginnings because the culture at one point lacked the information to explain things scientifically. In such myths, then, we find the origins of certain recognizable rites, of places and objects sacred to the culture. We also find more general explanations of the "How-the-Leopard-Got-Its-Spots" sort. In a more important sense, however, creation myths, and other myths, describe an understanding that is significant whether advanced science exists or not. A modern Hindu scientist might well subscribe to the "scientific" big bang theory of creation and in another part of his or her being have faith in the Hindu creation myth as a true metaphor for an ultimate reality that transcends science. A creation myth conveys a society's sense of its particular identity; it reveals the way the society sees itself in relation to the cosmos. It becomes, in effect, a symbolic model for the society's way of life, its world view—a model that is reflected in such other areas of experience as ritual, culture heroes, ethics, and even art and architecture.

The complex ceremony of the whirling dervishes of Islam provides mystical experience of the perfectly interrelated geometric universe created by the mysterious power that is Allah. The Navajo sand painting is a visual metaphor of creation, through contact with which the individual in need of curing can begin life again. In the old form of the Christian ceremony of the Eucharist, in which the communicant partakes of the sacred meal, the rite was concluded with a recitation of the Christian creation myth of John I (as opposed to the older Judaic myth of Genesis), indicating its deep structure as a curing ceremony of re-creation. The divine nature of the Christian culture hero, Jesus, is a reflection of the idea of the eternal Word or Logos that is revealed in John's creation story as the beginning of all that exists ("In the beginnning was the Word"). The ethics of Judaism are a logical derivation from the "just" and paternalistic creation expressed in Genesis. The Hopi underground kiva, where men weave and the most sacred ceremonies take place, is a microcosmic expression of the emergence creation story centered in the world womb of the ever-creatively-spinning Spider Woman.

While it is true that each creation myth reveals the priorities and concerns of a given culture, it is also true that when creation myths are compared, certain universal or archetypal patterns are discovered in them. Behind the many individual creation myths is a shadow myth that is the world culture's collective dream of differentiation (cosmos) in the face of the original and continually threatening disorder (chaos).

The basic creation story, then, is that of the process by which chaos becomes cosmos, no-thing becomes some-thing. In a real sense this is the only story we have to tell. Story-telling, like painting, singing, dancing, lovemaking, and eating, is a form of recreation, and it is well to remember that recreation has as its goal renewal or re-creation. The longing for this re-creation lies behind the painter's attempt to wrest significance from the resisting chaos of the blank canvas, behind the poet's struggle to convey meaning in overused words that long to become trite. It lies behind our attempts to "make something" of our lives, that is, to make a difference in spite of the seemingly universal drive toward meaninglessness or mere routine. In short, the archetype of the creation myth speaks to the equally universal drive for differentiation from nothingness that is expressed by everything that exists in the universe.

Creation myths convey the great struggle to exist in several basic symbolic structures about which, with some variations, scholars are in general agreement. Creation occurs primarily in one of five ways: 1) from chaos or nothingness (ex nihilo), 2) from a cosmic egg or primal maternal mound, 3) from world parents who are separated, 4) from a process of earth-diving, or 5) from several stages of emergence from other worlds. In every case there is a sense of birth—both of the world and of humans. Several of these structures can appear in a given creation myth.

A myth often attached to the creation myth is that of the Deluge. In this story the creator feels a mistake has been made or is disgusted with early humanity and clears the boards by sending creation back to the chaos of the Flood; although destructive of the old creation, the waters are also the maternal source for a new birth. They often support an ark or similar structure in which the seeds of a second creation are protectively contained.

Several archetypal characters appear consistently in creation myths. These include 1) a creator or the creatrix—the primal, ordered form that wrenches cosmos from chaos, sometimes from clay, sometimes from the fluids of its own body, and sometimes in conjunction with an equal and opposite natural power; 2) the trickster, who is sometimes a negative force and sometimes a culture hero who dives to the depths of nothingness to find form; 3) a first man and first woman, who continue the process of creation in our time and space and who sometimes fall from the creator's grace and are punished; and 4) the flood hero, who, floating in the placental ark, represents our never-ending urge for a new beginning.

All of the motifs, structures, and characters mentioned above will be confronted in some detail as particular creation myths are considered in this book. They are all elements of the symbolic dream language through which the human species fulfills an aspect of its role as the consciousness organ of creation.

Encyclopedia of Creation Myths

Acoma Creation

The Acoma Indians of New Mexico live in an ancient pueblo commonly called the Sky City, since it is perched atop a 600-foot maternal mound or butte. Like many of the Western American Pueblo cultures, Acoma society is matrilineal (ownership is passed down through the female line). Not surprisingly, then, the Acoma creation is orchestrated by a female spirit and is representative of the emergence type of creation myth, a birth process that begins in the earth womb (see also **Creation by Emergence**).

In the beginning two sister-spirits were born in the underground. Living in constant darkness they grew slowly and knew one another only by touch. For some time they were fed by a female spirit named Tsichtinako (see also **Spider Woman; Thinking Woman**) who taught them language.

When the proper time had arrived, the sisters were given baskets containing seeds for all the plants and models of all the animals that would be in the next world. Tsichtinako said they were from their father and that they were to be carried to the light of the upper world. She helped the sisters find the seeds of four trees in the baskets, and these seeds the sisters planted in the dark. After a long time the trees sprouted and one—a pine—grew sufficiently to break a small hole through the earth above and let in some light. With Tsichtinako's help, the spirits found the model of the badger, to whom they gave the gift of life and whom they instructed to dig around the hole so it would become bigger. They cautioned the animal not to enter into the world of light, and he obeyed. As a reward he was promised eventual happiness in the upper world. Next the sisters found the model of the locust in the baskets. After they gave him life, they asked him to smooth the opening above but warned him not to enter the world of light. When he returned after doing his job he admitted he had indeed passed through the hole. "What was it like up there?" the sisters asked. "Flat," he answered. Locust was told that for having done his work he could accompany the spirits to the upper world but that for his disobedience he would live in the ground and would have to die and be reborn each year.

Then it was time for the sister-spirits to emerge. Instructed by Tsichtinako, they took the baskets, Badger, and Locust, climbed the pine tree to the hole above, and broke through into the world. There they stood waiting until the sun appeared in what Tsichtinako had told them was the east. They had learned the other three directions from her, too, as well as a prayer to the sun, which they now recited, and the song of creation, which they sang for the first time.

Tsichtinako revealed that she had been sent to be their constant guide by the creator, Uchtsiti, who had made the world from a clot of his blood. The sisters were to complete the creation by giving life to the things in the baskets. This they did by planting the seeds and breathing life into the animals, but when the first night came the sisters were afraid and called on Tsichtinako, who explained that the dark time was for sleep and that the sun would return.

The creation was duly completed by the sisters, who took the names Iatiku (Life-Bringer) and Nautsiti (Full Basket). Some say the sisters quarreled and that Nautsiti disobeyed their father by giving birth to two sons fathered by hot rain drops from the rainbow and that, as punishment, the sisters were deserted by Tsichtinako.

It is said that one of the boys was brought up by Iatiku. When he was old enough, he became his aunt's husband. Together they made the people. Some say that Iatiku later created the spirits, or kachina, who would spend part of the year in the sacred mountains and part of the year with the people dancing for them in a way that would bring rain.
Sources: Erdoes, 97–105; Weigle, 215–218.

Adam

Adam is the first man of the Hebrew creation story. His name may mean red earth, suggesting his having been formed from clay (see also **Creation from Clay**). Adam was Eve's husband. Together they were to be the guardians of Yahweh's creation. Their disobeying of Yahweh led humanity's fall from grace (see also **Hebrew Creation**).

The Hebrew creation relates that Yahweh, or Elohim, made Adam, the first man. Eighteenth-century English poet and artist William Blake represented Elohim Creating Adam *in this watercolor.*

Ainu Creation

The Ainu, the original inhabitants of Japan who speak a non-Japanese language, live in the far northern islands of the country. Their mythology has characteristics in common with that of the Japanese, with whom they have interacted since the early ninth century. The Ainu creation story has characteristics of the earth-diver type (see also **Earth-Diver Creation**).

At first there was a mixture of mud and water and no living things. Then the Creator, who lives in the heavens above, sent a small bird, a wagtail, to make earth. Confused, the bird flew down and fluttered about and beat its wings at the mushy surface until a few dry spots emerged, forming the islands where the Ainu live.

As for the creation of humankind, for the most part the Ainu think that since they are so hairy themselves, their ancestor must be the polar bear.

A myth is also told of a heavenly couple sent by the Creator. They were called Okikurumi and Turesh, and they lived on a mountaintop. The couple had a son whom some call the first Ainu (*ainu* means human), and he is said to have taught the people how to survive. Still, most people say the first ancestor was the bear god (see also **Japanese Creation**).
Source: Leach, 205–207.

Alchemical Creation

In medieval times alchemy was a form of chemistry by which humans hoped to change base metals into valuable ones. In various cultures, however—including Chinese, Indian, Hellenistic, and modern Western—alchemical practices have had a spiritual aspect that represents the creative liberation of the world soul—or, in Jungian psychology, the individuated self. The use of alchemy as a metaphor for a new creation was particularly prevalent in the traditions of Gnosticism and Hermetism, which developed in Hellenistic Egypt from an amalgamation of ancient Egyptian religion, Greek philosophy, and Jewish and Christian mysticism. In this spiritual alchemy the soul was contained in the primal egg, the *prima materia,* which is the source of life in several creation myths (see also **Egyptian Creation; Gnostic Creation**).

Algonquin Creation

This earth-diver myth is similar to that of another Native American tribe, the Senecas. In this case the Sky Mother falls from the moon and produces a girl child who later mothers the culture hero Hiawatha (see also **Earth-Diver Creation; Seneca Creation**).

Altaic Creation

The Altaic tribes in Mongolia are influenced by their shamanistic past and their contact with major world religions such as Islam, Buddhism, Zoroastrianism, and Christianity. The creation myth of these peoples is particularly interesting in that it gives the devil a significant role in cosmogony by equating him with the first human. The myth is an earth-diver story (see also **Earth-Diver Creation**).

When there was only the primordial waters, two black geese flew back and forth over them. One was really God. The other was the first human; he was also the devil, however, and he could not resist trying to fly higher than God. Naturally God made him fall into the waters. When Man-Devil begged for help, God made him bring up a rock and then earth, which God turned into the world. When God asked Man-Devil to bring him more earth, Man-Devil did so, but he hid some in his mouth, thinking he would create his own world in secret when God was not looking. Both the earth that he handed God and the earth in his mouth began immediately to grow. Pained by the enormous swelling of his mouth, Man-Devil begged once again for God's help, and God chastized him before allowing him to spit out the material in his mouth, which became the earth's wetlands.

Another Altaic creation story, this one of the **creation from chaos** variety from Siberia, identifies the creator god as Ulgen and the first man as Erlik, who soon turns to evil ways and becomes, in effect, the devil in the myths of the region. Like the serpent in the Hebrew creation he corrupts the first woman (see also **Hebrew Creation**).

When Ulgen saw mud floating on the waters of pre-creation, he saw a human face in it and gave it life. Thus the first man, Erlik, was born (see also **Creation from Clay**). Soon Erlik forgot his place and boasted that he could create a man as well as Ulgen could. Ulgen reacted by flinging his first creation into the ends of the earth, where he reigns as the devil. Often Erlik returns to the upper world and brings evil with him.

After the fall of Erlik, God created the earth and placed eight trees and eight men on it. The eighth man, Maidere, and the eighth tree stood on a mountain of gold, and at Ulgen's bidding Maidere created the first woman.

When he saw that he could not give the woman life, Maidere left her in the care of a furless dog and went to get help from Ulgen. While he was away, Erlik came and offered the dog a fur coat in exchange for a look at the woman. He not only looked, but he also played seven flute notes into her ear, and she came to life possessed of seven tempers and many bad moods.

When Maidere returned he was surprised to find the woman alive. When he learned of Erlik's deception and the dog's betrayal, he condemned the animal to a life of bad treatment (see also **Buriat Creation; Creation Myths as Explanation; Samoyed Creation; Siberian-Tartar Creation**). ***Sources:*** Hamilton, 29–33; Sproul, 219–220.

Animism

Animism as a concept was first articulated by Sir Edward Taylor in his anthropological work, *Primitive Culture* (1871). The term *animism* is derived from the Latin word for soul, and the concept assumes the existence of a universal soul or spiritual power that is reflected in the existence of the spiritual aspect of all living things, especially in the existence of souls in human beings. Any creation myth that stresses the spiritual or godly essence of each element of the creation might be called animistic (see also **Chinese Creation; Okanagan Creation**).

Apache Creation

The Tinde (the People) were named Apache (the enemy) by the Pueblo people they raided. The Apaches are related to other Athabascan-speaking peoples such as certain Eskimos and the Navajos. They came to the Southwest relatively late compared to the Pueblo people—perhaps as late as 1000 C.E. The Apaches are now divided into five basic groupings, the White Mountain or Western Apaches in eastern Arizona, the Chiricahua in southwestern New Mexico (famous for the great warrior Geronimo), the Jicarilla in northeastern New Mexico, the Mescalero in southeastern New Mexico, and the Lipan in southeastern Texas. Not surprisingly, given the scattering of the tribe and its tendency to break up into still smaller subgroups, there are many Apache creation myths.

The Jicarilla people tell emergence creation stories (see also **Creation by Emergence**). These myths have clear connections to the idea of gestation and birth. In one myth, the underworld in which the people begin their existence is a great swelling womb. They enter the world by an opening at the top of a mountain after the waters of the earth have "broken." Appropriately, the creation myth is of great importance in the puberty ceremonies of Apache girls.

In the beginning the earth was only water and the people, the animals, and plants lived in the dark underworld. The darkness was pleasant to the animals we think of as night animals—the owl or the mountain lion, for instance—but not to the liking of the people and other day animals. Arguments ensued, and to settle them everyone agreed on a game to determine if there would be light or darkness. The game, still played today by Apache children, involved finding a button by looking through the thin wood of a thimblelike object. The day animals were better at this game and were rewarded with the rising of stars and then of the sun. When the sun got to the top of the underworld, he found a hole and saw the earth on the other side. When he told the people of the world above, they all wanted to go there, so they built four mounds—one for each direction—and planted them with various fruits and flowers. Both the mounds and the plantings grew until two girls climbed them to pick flowers; all growth then stopped, leaving the mountaintops still far from the hole above.

It took the help of the buffalo to get the people up to the hole. They contributed their long straight horns to be used as a ladder, and it is because of the weight of the climbers that the buffalo's horns to this day are curved. Before they emerged from the hole the people sent up the moon and sun to provide light and four winds—one from each underground mound—to blow away the waters that covered everything. After various animals had gone through the hole to test the new world, the people emerged and travelled in each of the four directions until they reached the seas. On these journeys the individual tribes broke off to make their homelands. Only the Jicarilla Apaches stayed behind, constantly circling the hole from which they had come, and eventually the Great Spirit settled them there in what is the center of the world.

Another Jicarilla emergence myth, with **creation from chaos** and animistic characteristics, gives a prominent role to kachinalike personifications of the basic natural powers (see also **Animism**). These beings, called Hactcin, existed before creation, when there was only dark, wet chaos—the world womb, as it were. Being lonely, the Hactcin created the essential elements of the universe and also created Earth Mother and Sky Father. As for the people, at this time they lived only as potential form in the damp dark underworld, where a figure called Black Hactcin ruled. Black Hactcin was the true creator. He joyfully made animals out of clay and then taught them how to reproduce themselves and what and how to eat. Then he told them to find appropriate places to live, and they did—the buffalo went to the plains, certain sheep and goats to the mountains, prairie dogs under the ground, birds to the air and trees, crickets to the grass, frogs and fish to the water, and so forth. Black Hactcin also called down water from the sky, and he invented seeds.

The animals asked him to give them a special companion, one who could take his place in case he ever decided to hide from the world. Black Hactcin agreed and began work on the creation of humankind. The animals helped him by gathering the essential materials—pollen, clay, valuable stones, minerals—but Black Hactcin made them stay away from him while he worked. After facing the four directions, he made a sketch of his own shape on the ground. Then he used the gifts of the animals to flesh out the various parts of his creation. For instance, red ochre was used for blood, coral for skin, rock for bones, opal for fingernails and teeth, and abalone for the white part of the eyes. Black Hactcin used a dark cloud for the hair, and of course the cloud would later become white. To bring the first man to life, Black Hactcin blew wind into him. Then he raised him up and commanded him to speak and then to laugh, shout, and walk. Black Hactcin also made the man run in a certain pattern, which is why, at a girl's puberty ceremony today, the girl must run in the same pattern.

First Man lived alone with the animals, and they all spoke the same language. The animals told Black Hactcin that the man needed a personal companion, however, and the creator made him dream of a woman, who was there with him when he woke up. First Man and First Woman were happy.

A variation of the Jicarilla creation says that it was the dog that asked the creator for a companion and that it was he who drew the sketch of man with his paw. When the man blossomed into life, he and the dog went off together like best friends.

The Mescalero Apaches emphasize the connection between a girl's puberty rites and the creation. When a girl menstruates for the first time, a sacred lodge is built for her; the form of the lodge is based on the created universe—a circle bisected along the four directions. Of the twelve poles holding up the lodge, the four main ones are the Four Grandfathers who hold up the universe. They are the four directions, the four seasons. The puberty ceremony itself lasts four days and four nights and is, of course, a re-creation of the first human in the newly blossomed woman. The *ex nihilo* creation myth is recited at the ceremony (see also **Creation from Nothing**).

Once there was only the Great Spirit. He created the world in four days. He made Father Sun, Mother Earth, Old Man Thunder, Boy Lightning, and the animals. Then on the fourth day he made the People, the Tinde.

The following is a White Mountain Apache myth of the Four Grandfathers and the sacred lodge. In it the universe itself is the lodge. An animistic story of the **creation from chaos** type, this version was told by a very old shaman, or medicine man, named Palmer Valor, on a winter night in 1932 (see also **Animism**). The Apaches, like many other Native Americans, believe that sacred stories must be told only at night and only in the cold months. In the daytime or during warmer months, dangerous beings—snakes, scorpions, lightning—might be able to hear themselves talked about and then might punish both the storytellers and the listeners.

> Four people started to work on the earth. When they set it up, the wind blew it off again. It was weak like an old woman. They talked together about the earth among themselves. "What shall we do about this earth, my friends? We don't know what to do about it." Then one person said, "Pull it from four different sides." They did this, and the piece they pulled out on each side they made like a foot. After they did this the earth stood all right. Then on the east side of the earth they put big black cane, covered with black metal thorns. On the south side of the earth they put big blue cane covered with blue metal thorns. Then on the west side of the earth they put big yellow cane covered with yellow metal thorns. Then on the north side of the earth they put big white cane covered with white metal thorns.
>
> After they did this the earth was almost steady, but it was still soft and mixed with water. It moved back and forth. After they had worked on the earth this way Black Wind Old Man [the wind of the east] came to this place. He threw himself against the earth. The earth was strong now and it did not move. Then Black Water Old Man threw himself against the earth. When he threw himself against the earth, thunder started in the four directions. Now the earth was steady, and it was as if born already.
>
> But the earth was shivering. They talked about it: "My friends, what's the matter with this earth? It is cold and freezing. We better give it some hair." Then they started to make hair on the earth. They made all these grasses and bushes and trees to grow on the earth. This is its hair.
>
> But the earth was still too weak. They started to talk about it: "My friends, let's make bones for the earth." This way they made rocky mountains and rocks sticking out of the earth. These are the earth's bones.

Then they talked about the earth again: "How will it breathe, this earth?" Then came Black Thunder to that place, and he gave the earth veins. He whipped the earth with lightning and made water start to come out. For this reason all the water runs to the west. This way the earth's head lies to the east, and its water goes to the west.

They made the sun so it traveled close over the earth from east to west. They made the sun too close to the earth and it got too hot. The people living on it were crawling around, because it was too hot. Then they talked about it: "My friends, we might as well set the sun a little further off. It is too close." So they moved the sun a little higher. But it was still too close to the earth and too hot. They talked about it again. "The sun is too close to the earth, so we better move it back." Then they moved it a little higher up. Now it was all right. This last place they set the sun is just where it is now.

Then they set the moon so it traveled close over the earth from east to west. The moon was too close to the earth and it was like daytime at night. Then they talked about it: "My friends, we better move the moon back, it is like day." So they moved it back a way, but it was still like daylight. They talked about it again: "It is no good this way, we better move the moon higher up." So they moved it higher up, but it was still a little light. They talked about it again and moved it a little further away. Now it was just right, and that is the way the moon is today. It was night time.

This is the way they made the earth for us. This is the way all these wild fruits and foods were raised for us, and this is why we have to use them because they grow here.

Reprinted from Granville Goodwin, *Myths and Tales of the White Mountain Apache*. Memoirs of the Apache Society, vol. 33, New York: J. J. Augustin, for The American Folklore Society, 1939.

A creation myth of the Chiricahua Apaches is influenced by contact with the white invaders of their land. Strictly speaking, it is a re-creative flood myth rather than a creation myth.

The first people on earth did not know anything about the Great Spirit. They only knew the Hactcin, the spirits of the earth, who lived in the mountains. The Great Spirit was not pleased, so he sent the Flood and most of the world perished. Some of the people and animals saved themselves by climbing White-ringed Mountain (near what is now Deming, New Mexico). The turkey was the last one up, and he got his tail feathers wet, which is why they are tipped with white today. When the waters withdrew, the saved people and animals went down the mountain and something strange happened. Two men were made to stand before a gun and a bow and arrow and were told to choose between them. The one who chose first took the gun and became the White Man; the one who got the bow and arrow became the Indian.

Like several other emergence myths the Lipan emergence creation clearly suggests an analogy with birth (see also **Acoma Creation; Creation by Emergence**). Everything begins in the womb of Earth herself. Many of the characters in this Apache myth—Killer of Enemies and Changing Woman, for example—are to be found also in the mythology of another Athabascan tribe, the Navajos.

When the people lived in the lower world in darkness, they wondered if there was a different kind of world anywhere else. It was decided in council that someone should be sent out to explore, and Wind agreed to go. He went up to our world and blew away some of the waters that covered everything, and there was land. He did not go back to the people below the way he had promised, though. The people then sent Crow out and he did not come back either; instead, he stayed and picked the eyes out of the dead fish he found on the new land.

Finally the people sent out Beaver, but he amused himself by building dams in the streams he found trickling through the new land, and he, too, failed to come back.

It was only when they sent faithful Badger that the people found out about our world. Badger went up, looked around, and came back to report on everything he had seen.

Finally, the people sent up the Four Grandfathers; they were the first Indians, and they arranged the world for us. They did this by turning one of themselves into a huge ball. Out of this ball they fashioned the trees, mountains, and streams. When everything was as it should be, the people below were called, and they came out into this world.

After the emergence, the people wandered about, and some stopped at various places, forming the tribes. They were led by Killer of Enemies (Sun) and Changing Woman (Moon). The last people to settle were the Lipan Tinde. Sun and Moon vowed to separate from the people and from each other and to keep on moving (see also **Creation from Division of Primordial Unity**). They would meet each other once in a while in eclipses.

Sources: Erdoes, 83–85; Leach, 72–74; Sproul, 258, 260–267; Williamson, 304.

Arandan Creation

This myth from northern Australia, an example of **creation from chaos** with elements of the **creation by emergence** theme, contains the familiar elements of **creation by thought** or **the Dreaming** and the creation that has gone wrong and must be cleansed by a flood (see also **The Flood and Flood Hero; Imperfect Creation**). In the myth's mentioning of places and objects familiar to the Arandan aborigines, we find the common understanding on the part of a given culture that creation began in the center of its local world. The ratlike bandicoot, for example, is sacred to the Bandicoot clan, whose specific creation myth this is, and Ilbalintja soak is a real place. The decorated *tnatantja* pole—a kind of axle tree or world tree (see also **Yggdrasil**) and the bull-roarer are still used in Arandan religious ceremonies. The bull-roarer is an object found in many societies; it is a flat piece of wood with pointed ends. At one end is a hole through which a hair string is attached so the bull-roarer may be spun around to make a mysterious buzzing sound.

In the beginning when there was darkness everywhere, the creator, Karora, lay sleeping in Ilbalintja, covered by rich soil and a myriad of flowers and other plants. From the center of the ground above Karora rose a magnificently decorated and living tnatantja pole that reached all the way to the sky. Under the ground the god's head lay on the roots of the pole, and in his head were thoughts that became real. Huge bandicoots slithered out from his navel and his armpits—male wombs—and broke through the soil above, and the sun began to rise over Ilbalintja.

The sun having brought light, Karora burst through the earth, leaving a gaping hole—Ibalintja soak—which filled with the bloodlike juice of the huneysuckle. Having left the earth, Karora's body lost its magical powers, and the god became hungry. He grasped two of the bandicoots writhing around him and roasted them in the heat of the new sun.

As the sun went down decked in necklaces and a veil of hair strings, the great ancestor thought about a helper but fell asleep with his arms stretched out. As he slept, a bull-roarer emerged from his armpit and turned into a young man, Karora's first son. In the morning Karora woke up to find his new companion lying next to him but without life. The ancestor, his body now decorated, made the sacred *raiankintja* call. The sound gave life to his child, and father and son did the ceremonial dance.

During the next nights Karora gave birth to many more sons, all of whom became hungry and ate bandicoots until none was left. Karora sent his sons into the plains to find more bandicoots, but they returned hungry.

On the third day the sons heard what they thought was a bull-roarer sound and began searching in bandicoot nests until a strange hairy animal hopped out. "It's a sandhill wallaby," the men shouted, and they broke one of the animal's legs with their sticks before it could cry out, "You have lamed me; I am not a bandicoot, but Tjenterama, a man like you." The hunters backed off as the kangaroo limped away.

Karora met his sons when they returned home. He led them to Ilbalintja soak and ordered them to sit in a circle around it. Then the honeysuckle juice rose and swept them down into the soak and underground to the injured Tjenterama, their new leader. They remained there forever and became objects of worship to the people who came later.

Karora returned to his old sleeping place in the soak. The people still go there to drink and honor him with gifts of greens. Karora smiles in his sleep, happy to have them visit.

Source: Hamilton, 47–51.

Arapaho Creation

The Arapaho Indians of the Great American Plains migrated westward from the eastern woodlands. So it is that their creation myth is of the **earth-diver creation** type so common among the Indians of what is now the eastern United States. It is also an example of **creation by thought.** The Arapaho creator is a personified version of the flat pipe so important in their ceremonies. The Great Spirit is a formless projection of his thoughts.

In the beginning there was water everywhere and on it floated Flat Pipe all alone. The Great Spirit called down to Flat Pipe suggesting that he create beings to help him build a world around him. Flat Pipe thought of ducks, and they appeared. He ordered them to dive below the water's surface to see what was there, but they could not reach the bottom. The same thing happened when Flat Pipe created other water birds.

Finally, the Great Spirit made him think of an animal that could live in water or on land, though Flat Pipe had to conceive of land before there could be any. The animal he thought of was the turtle, who agreed to dive into the waters to find land. After a long time, Turtle returned and she spit out a piece of land onto Flat Pipe. Out of this land grew the earth as we know it, and out of it Flat Pipe made man and woman and all the animals, and they multiplied (see also **Creation from Clay**).

Source: Marriott (B), 27–29.

Arikara Creation

The Arikara are Plains Indians whose land is adjacent to that of the Mandans, and some versions of their creation myth are clearly related to the **Mandan creation** myth. The familiar figure of Lone Man is present in one version. He is born of a plant and seems to have developed under the influence of the Christian missionaries.

The primary Arikara creation—an emergence myth (see also **Creation by Emergence**)—is recited at the spring ceremonies to celebrate the opening of Mother Corn's "sacred bundle."

In the beginning the great sky chief, Nishanu, made giants, but these creatures had no respect for their maker and were destroyed by a great flood (see also **The Flood and Flood Hero**). Only a few good giants were preserved as corn kernels under the ground. Nishanu also planted some corn in the heavens. Out of this corn came Mother Corn, who descended to the earth to lead the people out. Since the people were still animals then, they dug their way out with Mother Corn's encouragement. Then the mother led the people from the east, where they had emerged, to the west, where they are now.

Mother Corn then went back to heaven, but while she was gone the people made trouble and started killing each other. She returned later with a leader for the people, named Nishanu after his maker, in whose image he was made. The leader taught the people how to fight enemies rather than each other. Mother Corn taught them the ceremonies.

Sources: Bierhorst, 166; Sproul, 248.

Assiniboine Indian Creation

This Native American **earth-diver creation** is dominated by the sometimes **trickster** creator, Inktonmi, or Inktome or Inktomi, who is often thought of as a spider among other Plains Indians (see also **Sioux Creation**). The myth as it has come down to us has had important elements of tribal life added to it in order to give those elements sacredness. The presence of horses in the myth (even though these animals did not come to North America until the Spanish brought them) is a good example.

When everything was water, Inktonmi sent various animals to find earth below the primeval sea. Only the muskrat succeeded; he floated up dead but there was earth in his claws, and out of that earth the creator made land. He then said there would be as many winter months as there were hairs in his fur robe. Only the frog dared point out to Inktonmi that this

would be too many months of winter and suggested that seven cold months would be sufficient. When he continued to argue his point, Inktonmi killed him, but even after death he signified seven months with his toes, and the the creator gave in to the frog's idea. Finally, Inktonmi made people and horses out of dirt, and he taught the Assiniboine how to steal horses from other peoples.
Source: Sproul, 252–253.

Assyrian Creation

The Assyrian Empire of ancient Mesopotamia achieved its greatest size in the seventh century B.C.E. The Assyrians go back to much earlier times, however. With the Sumerians, Babylonians, and Egyptians, they created the amazingly advanced civilization of the ancient Near East, and not surprisingly, elements of each other's religions are to be found in all of those cultures. The Assyrians spoke Akkadian, a Semitic language that was also the language of the great Babylonian epic, the **Enuma Elish** (see also **Babylonian Creation; Egyptian Creation; Sumerian Creation**). The Assyrian capital was at Ashur and later at Nineveh. Assyrian creation stories vary greatly from period to period, depending in part on the power of various deities at any given time.

In an Assyrian creation used for religious initiation ceremonies, we find a pantheon of dominant male gods, but there are fragments in the myth of earlier *ex nihilo* creations suggesting a dominant goddess (see also **Creation from Nothing**).

After the earth and heavens had been created and the Mother Goddess, too, the great sky gods—the Annunaki—led by Anu (sky), Enlil (storms and earth), Shamash (sun), and Ea (water), looked out over their creation and wondered what else they needed to do. The beautiful Tigris and Euphrates rivers flowed majestically to the sea, and the destinies of heaven and earth were established, but something seemed to be lacking. It was decided that mankind was needed to till the fields, celebrate religious festivals, and constantly retell the origin stories. The new being would be made of the blood of certain sacrificed deities (see also **Creation by Sacrifice**). So it was that the first humans—Ulligarra (abundance) and Zalgarra (plenty)—were created. Their destinies were established by the "lady of the gods," Aruru.

In another Assyrian myth, however, it is the goddess herself, Ninhursag (also Nintu or Mama, goddess of earth) who creates the human. This myth was apparently used as part of a birth incantation. The myth itself depicts the birth process.

After the great goddess is praised and her feet kissed, she goes with the other gods to the House of Fate, where 14 "mother-wombs" (pregnant women in the ritual) are assembled. The great god Ea sits next to the goddess and asks her to begin the incantation. She does so, drawing 14 figures in the clay before her and then pinching off 14 pieces, placing seven to her left and seven to her right with a brick between them. Then Ea kneels on a mat, opens his navel, and calls on the mother-wombs to bring forth seven males and seven females. Then the Great Mother Womb, Ninhursag, herself forms the new beings.

During the incantation itself, the mother in the birthing house is encouraged to act for herself as the goddess and to bring forth her child safely.
Source: Sproul, 115–116, 118–120.

Atum

Atum is a version of the Egyptian god who creates the world via masturbation (see also **Creation by Secretion; Egyptian Creation**).

Australian Aborigine Creation

Several origin myths belong to many of the aboriginal people. The Arandans, the Murngins, and the Great Western Desert people, for instance, tell of a prehuman "dream time" during which magical ancestors created sites, traditions, and people during their "walkabouts" (see also **Djanggawul Creation; The Dreaming; Ngurunderi Creation**). Often the ancestors became lizards (or other animals), gave birth to more lizards, and then, warmed by the sun, turned as a group into humans. The Dieri god made the first man in the form of a lizard but found it could only walk when its tail was cut off. The Arandan people have a creator-lizard, Mangw-er-kunger-kunja, and the Jumu and Pindupis have one called Pupola. The creator-lizards gave the bull-roarer and other sacred totem objects to their people. They also gave them the boomerang and taught them how to survive (see also **Arandan Creation; Djanggawul Creation**).

Aymaran Creation

The Aymara are Andean people of Bolivia. Their principle deity is the snow god, Kun. Angry at human beings, Kun once covered creation

The Gunwinggu people of Australia's Arnhem Land attribute their creation to Waramurungundi, the first woman, shown here carrying a digging stick and a string bag, and to Wuragag, her husband, shown on the next page.

Wuragag, husband of Waramurungundi, the Australian aborigine creator, shown on the previous page, is shown here carrying a spear and wearing feathers in his hair.

with snow and ice, and nothing but evil spirits could survive on the frozen world. After this "ice flood," it was the gods of fertility who sent their sons, the Eagle Men, to create new people, the Paka-Jakes, who still live near Lake Titicaca.
Source: Freund, 6.

Aztec Creation

In this world-parent separation myth, it is said that two gods, Quetzalcoatl (the Plumed Serpent) and Tezcatlipoca, pulled the earth goddess, Coatlicue (Lady of the Serpent Skirt), down from the heavens, took the form of great serpents, and ripped her into two pieces to form an animistic earth and sky (see also **Animism; Egyptian Creation; Geb and Nut; World Parent Creation**).

In her earthly form, Coatlicue was the source of all nature. Her hair became plants; her eyes and mouth were caves and sources of water. Other parts of her body became mountains and valleys. Coatlicue, however, was angry at what had been done to her, and this is why she often demanded human hearts and human blood.
Source: Weigle, 55–56.

The Aztec of fifteenth-century Mexico depicted the birth of gods, including Quetzalcoatl, the Plumed Serpent, center, on a page of this painted book. Two oblong blades, like the blades used to cut out the hearts of living human sacrificial victims, form the mother's head.

Babylonian Creation

The Babylonians lived in the so-called Fertile Crescent, the valley of the Tigris and Euphrates rivers in what is now Iraq. Their great civilization followed that of the Sumerians, which influenced their culture. These cultures brought high civilization to Mesopotamia several millennia B.C.E. (see also **Assyrian Creation; Sumerian Creation**). The best known of the Babylonian creation stories is contained in the **Enuma Elish,** the great epic named after its first two words, *enuma elish,* meaning "when on high."

The Enuma Elish is a world-parent type of creation myth (see also **World Parent Creation**). Recorded in the form we have come to know it best in about 1100 B.C.E. during a celebration of Nebuchadnezzar's recapture of the city's statue of Marduk, the poem is written in a Semitic language, Akkadian—the language of Mesopotamia in the third millennium B.C.E. Parts of the story have been found in cuneiform script on clay tablets dating from about 2500 B.C.E., and it seems more than likely that it is based in part on earlier Sumerian texts, especially since many of the gods mentioned are of Sumerian origin. However we date it, the Enuma Elish is one of our oldest extant creation stories, and it is one of the most famous.

The work is an unusual creation epic for the polytheistic world of the ancient Near East in that Marduk assumes many of the functions and responsibilities traditionally given to other gods in the local pantheon. It is also quite unlike the monotheistic biblical Genesis in that it concentrates on the process of creation rather than its results. Creation for the Babylonians is a form of procreation or of craftsmanship (see also **Creation by *Deus Faber***).

Cuneiform inscriptions relate the Babylonian creation by the god Marduk on this tenth-century B.C. clay tablet.

The Enuma Elish served the Babylonians well. Living in an unpredictable river valley adjacent to the Persian Gulf, they would recite yearly the story of the battle between Marduk and the watery chaos (Tiamat) during the New Year's rituals in Marduk's temple, rituals that coincided with the spring inundations. It should be noted that despite the natural surroundings of Babylon and traditional elements of creation perhaps adopted from the Sumerian cosmology, a dynamic is initiated in the Enuma Elish that would lead one to believe this story is a timely propaganda piece, written to justify Marduk's glory and control over the natural forces of the universe and Babylon's supremacy over the Fertile Crescent region.

Apsu, the primordial freshwater ocean "commingled" with the saltwater of Tiamat to bear Lahmu and Lahamu, silt deposits that eventually formed land. Out of the union of Lahmu and Lahamu also came the gods Anshar and Kishar and then their son, Anu, who fathered the mighty Ea and his brothers. Ea and his brothers roamed back and forth on the waters, and this bothered Tiamat and Apsu, the primordial parents. Apsu decided to act against the young gods, but they got wind of his plans and killed him.

Now the wise Ea and the goddess Damkina produced the great god Marduk. "My son, the Great Sun," Ea called him.

Meanwhile, Tiamat created monsters in her anger against the gods for having destroyed her consort and for having made the winds that disturbed her great body. Anshar, Anu, and Ea attempted to subdue Tiamat and her squadron of monsters but were unable to do so. It was then that Marduk made his move towards supremacy. He would conquer Tiamat if the gods would recognize him as king of the universe. After a test of his powers over the sky, they so recognized him, and Marduk prepared for battle.

Taking the form of warrior, Marduk took up his thunderbolt and rode to the now stirred-up waters of Tiamat, who became like a huge monster—some say a dragon. The god defeated the first mother, the primordial goddess, and cut her in half to form heaven and earth: "He stilled himself to observe the corpse of Tiamat,/ . . . He divided her like a shell-fish into two parts:/He threw one half to the heavens and called it the sky . . . /he formed the firmament below" (see also **Creation from Dismemberment of Primordial Being**).

Now began the process of turning what had been a natural but chaotic creation (beginning with the commingling of Apsu and Tiamat) into an active but ordered process. In Tiamat's "stomach he made the sun's path./He made the moon shine and gave her the night to hold." After creating the earth, the constellations, and positions and responsibilities for the gods, Marduk had Ea craft humans from the bones and blood of Tiamat's lover, the chief monster Kingu. These new beings were to serve the gods, especially in the sacred city of Babylon, where Marduk (king of kings, he of the fifty names) and his fellow gods would have their sanctuary.

The following is a section from the Enuma Elish as translated by E. A. Speiser.

> When on high the heaven had not been named,
> Firm ground below had not been called by name,
> Naught but primordial Apsu, their begetter,
> (And) Mummu-Tiamat, she who bore them all,
> Their waters commingling as a single body;
> No reed hut had been matted, no marsh land had appeared,
> When no gods whatever had been brought into being,
> Uncalled by name, their destinies undetermined—
> Then it was that the gods were formed within them.

Reprinted from a translation by E. A. Speiser in J. B. Pritchard, ed., *Ancient Near East Texts Relating to the Old Testament,* Princeton, NJ: Princeton University Press, 1955.

Another Babylonian myth resembles the Assyrian Mother Goddess creation (see also **Assyrian Creation**).The presence of the god Enki (the Sumerian name for Ea, god of waters) suggests an earlier source in pre-Semitic Sumerian mythology.

The goddess Mami (Mama, Ninhursag, Nunti) is called upon to create the first human, Lullu (the savage). "Let him be formed of clay, made alive by blood," she cries. The god Enki (Ea) supports Mami's demand and adds that the new being must be made of clay and the blood of a sacrificed god (see also **Creation by Sacrifice**). God and man would thus be joined "unto eternity." The goddess finally gives birth to the being she has designed with Enki's help.

What follows is a childbirth incantation similar to that found in the **Assyrian creation** and almost certainly in keeping with earlier Sumerian practice.

> "That which is slight shall grow to abundance;
> The burden of creation man shall bear!"
> The goddess they called, [. . .], [the *mot*]*her,*
> The most helpful of the gods, the wise Mami:
> "Thou art the mother-womb,
> The one who creates mankind.
> Create, then Lullu and let him bear the yoke!
> The yoke he shall bear, . . . [. . .];

The *burden* of creation man shall bear!"
.[.].opened her mouth,
Saying to the great gods:
"With me is the *doing* of all that is suitable;
With his . . . let Lullu appear!
He who shall be [. . .] of all [. . .],
Let him *be formed* out of clay, be *animated* with blood!"
Enki opened his mouth,
Saying to the great gods:
"On the . . . and [. . .] of the month
The purification of the land . . . !
Let them slay one god,
And let the gods be purified in the *judgment.*
With his flesh and his blood
Let Ninhursag mix clay.
God and man
Shall [. . .] therein, . . . in the clay!
Unto eternity [. . .] we shall hear."
[*remainder of obverse too fragmentary for translation*]
[*reverse*]
[. . .] her breast,
[. . .] the beard,
[. . .] the cheek of the man.
[. . .] and the raising
[. . .] of both eyes, the wife and her husband.
[Fourteen mother]-wombs were assembled
[Before] Nintu.
[At the ti]me of the new moon
[To the House] of Fates they called the *votaries.*
[*Enkidu* . . .] came and
[*Kneel*]*ed down,* opening the womb.
[. . .] . . . and happy was his countenance.
[. . . bent] the knees [. .],
[. .] made an opening,
She brought forth issue,
Praying.
Fashion a clay brick into a core,
Make . . . stone in the midst of [. . .];
Let the vexed rejoice in the house of the one in travail!
As the Bearing One gives birth,
May the mo[ther of the ch]ild bring forth by herself!

[*remainder too fragmentary for translation*]

In keeping with the Babylonians' tendency to use myths for ritual purposes, the following creation story was used for the dedication of sacred buildings. It has some resemblance to the architectural aspects so prevalent in Native American myths (see also **Apache Creation**).

In the beginning Anu (sky) created the heavens, and Nudimmud (Eawaters) created Apsu, the primeval waters. Then Ea took some clay and created necessary elements for the building of great structures. He made the Arazu, the gods of the various crafts—brick-making, carpentry, and so forth. Then he made the mountains and waters. He also made Kusiga, a master of ceremonies for the gods, and a king to maintain the temples. Man was made to serve the gods.

The architectural sense is preserved in a late Babylonian creation myth of the sixth century B.C.E. Marduk, the strong god of the Enuma Elish, is the dominant figure in that myth, which is associated with the city of Eridu in the marshlands at the north of the Persian Gulf. The myth was used for purification ceremonies at certain temples.

Before anything was built—before the cities of Eridu, Nippur, or Erech were built, even before the dwelling place of Ea, Apsu, or any temple had been made—all the lands were under the sea. Then came Marduk to make Eridu, Babylon, and the Anunnaki (gods).

Marduk covered some of the waters with a great reed frame and filled it with earth to make a place for the gods and mankind. He created mankind with the goddess Aruru, and he created the animals, the Tigris and Euphrates rivers, and the fields and forests, giving them all names. He dammed up part of the sea and built Eridu, creating people to live in it. Finally he built Nippur and the other cities of Babylon and made their people.

A very late Babylonian creation myth was written by Berossus, a priest of Marduk in about 250 B.C.E. Berossus has combined many familiar forms from the Enuma Elish and other Mesopotamian creations: the primordial waters as a goddess, the dominance of Bel (Lord) Marduk, the sacrificing or division of the first being, and the use of clay and sacrificed blood to create mankind. The presence of such beings as centaurs and satyrs suggests a Greek influence.

In the beginning there was only darkness and water. Out of this chaos sprang an army of oddly formed creatures: men with wings, two faces, or both; beings that were at once male and female; humans with goat feet; others who were part horse and part man. These creatures, all depicted in the temple of Marduk, were ruled over by Omorka, the moon (female).

Marduk cut Omorka (a version of Tiamat) in two, made one half into the sky and the other into the earth, and destroyed all of the monsters that had lived within her being. Then the world was empty, so Marduk commanded one of the gods to cut off his head, and from the blood and bits of earth he created the world—humans, animals, stars, sun, moon—everything that is (see also **Creation by Sacrifice**).

Sources: Brandon, 66–117; Kramer (B), 142–146; O'Brien, 10–32; Sproul, 91–117, 120–122.

Bagobo Creation

In their *ex nihilo* myth (see also **Creation from Nothing**), the Bagobo (or Bagopo) people of Mindanao Island in the Philippines say that in the beginning there was only the Creator, Melu. He lived in the heavens. He was white, he constantly polished his whiteness, and he had gold teeth. With the dried skin that came from his polishing, Melu made the earth and, in his image, two small people. The only thing lacking on each of the people was the nose. So Melu's brother offered to make them, and Melu only reluctantly allowed him to do so. The brother was not very smart, and he made the noses upside down.

When it rained for the first time, the first people almost drowned, until they stood on their heads under a tree. Melu came along and asked them what they were doing. Then he saw the upside-down noses and turned them around. Everything has been all right since—with noses.

Source: Leach, 164–165.

Bank Islands Creation

The Bank Islands are north of the Melanesian New Hebrides. The people there are ruled over by the sun god Quat, one of whose major tasks in the creation was to discover darkness, thus reversing the usual pattern of **creation from chaos** in which life begins in darkness and light is established later.

Once there was light everywhere all the time. The light shone on the mother-stone, Quatgoro, and one day she broke open to release Quat and his 11 brothers, all named Tangaro and each representing a characteristic or a plant. The brothers grew up immediately.

Quat carved the first humans from different parts of a tree, and then he pieced them together into puppetlike figures. When he had six of these puppets, he lined them up and danced in front of them until they began to come to life. Then he beat his sacred drum and they began to dance. Finally, Quat made six of the puppets into men and six into women, and they became mates.

Tangaro, the Foolish One, thought he could do what his older brother Quat had done. Using a different kind of tree, he carved six puppets and danced them into life, but then he buried them and forgot about them for a while. When he came back and dug them up they were dead and rotten, and so it is that we have death in the world.

As for Quat, he continued with his creative work. When he made pigs that walked on two legs, his brothers made fun of him, saying his pigs were too much like humans and silly looking. Quat shortened the pigs' front legs and they began to walk on all fours.

Quat made everything: canoes, plants, animals, rivers, and so forth. Then the brothers began complaining about all the light, so Quat got into his canoe and paddled to the edge of the world to Oong (night). Oong was completely dark and without light. It taught Quat about sleep and gave him dark eyebrows and a piece of itself to take back to his own world.

On his way home Quat stopped at the Torres Islands to exchange a bit of night for some birds. From then on birds have always followed night with their chirping so we are ready for day when it comes.

At home his brothers were waiting, and Quat taught them about beds (made of coco leaves) and sleep and made them lie down. Then he released bits of night, and the sun began to disappear. "What is happening?" the brothers asked. Quat comforted them and told them to be quiet. Soon they began drifting off to sleep and they were frightened; maybe they were dying. Quat reassured them; "It's only sleep," he said, and they became quiet.

While the Tangaro were sleeping, Quat cut a little hole in night with a sharp, red stone. When the birds welcomed the light and woke the brothers, they saw the red sunrise for the first time and were very happy. They began their day's work. So it still happens.

Source: Hamilton, 9–13.

Bantu Creation

See Boshongo Creation; Fang (Fans) Creation.

Barotse Creation

See Lozi Creation.

Bible

The Bible is a collection of books written over a period of some thousand years. The older part—the Old Testament—is made up of the Jewish scriptures; it includes the five books of the Torah (the Pentateuch), the books of the prophets, and the later Writings. The newer part—the New Testament—consists of the purely Christian scriptures, including the gospels (versions of the life of Jesus), the letters of Paul and other church leaders to the early Christians, and the prophetic and apocalyptic Book of Revelation. The traditional Judeo-Christian creation myth is contained in Genesis, the first book of the Bible (see illustration on page 32); a later **Christian creation** account is found in the first chapter of the Gospel of John (see also **Hebrew Creation**).

Big Bang Theory

Myths are considered truth by the cultures from which they first emerge—at least until they are "exposed" as "mere myth." The big bang theory, the currently accepted creation story of our scientific culture, reflects our cultural priorities; it is a record of our culture's understanding of its own place in the universe and its sense of what the universe is. It depicts a world created in a few minutes in one great explosion long, long ago. According to the theory, our solar system was organized by that explosion and has been expanding ever since. At this moment, we can see the moment of creation because the light from the first explosion reaches us now after a voyage taking 20 billion years. The big bang theory suggests that everything that exists has a common ancestry in a single primeval event (see also **Creation from Nothing; Creation in Science**).

Birth as Creation Metaphor

The birth of a child can be seen as a metaphor for or microcosmic representation of creation. Thus in some cultures the creation myth is recited to infants when they "enter the world." When an Osage Indian child

Incipit liber bresith q̄ nos genesī di-
In principio creauit deus celū
et terram. Terra autem erat inanis et
vacua: ⁊ tenebre erāt super faciē abissi:
et spiritus dñi ferebatur super aquas.
Dixitq; deus. Fiat lux. Et facta ē lux.
Et vidit deus lucem q̄ esset bona: et
diuisit lucem a tenebris. appellauitq;
lucem diem ⁊ tenebras noctem. Factū
q; ē vespere ⁊ mane dies unus. Dixit
quoq; deus. Fiat firmamentū in me-
dio aquarū: et diuidat aquas ab a-
quis. Et fecit deus firmamentū: diui-
sitq; aquas que erant sub firmamen-
to ab hijs que erant super firmamen-
tum: ⁊ factum est ita. Vocauitq; deus
firmamentū celū: ⁊ factum est vespere
et mane dies secundus. Dixit vero deus. Congregentur aque que sub celo
sunt in locum unū et appareat arida.
Et factum est ita. Et vocauit deus ari-
dam terram: cōgregationesq; aquarū
appellauit maria. Et vidit deus q̄ esset bonū. et ait. Germinet terra herbā
virentem et facientem semen: et lignū
pomiferū faciens fructum iuxta genus
suū: cuius semen in semetipso sit super
terram. Et factum est ita. Et protulit
terra herbam virentem et facientem semen iuxta genus suū: lignūq; faciens
fructū ⁊ habēs unūquodq; sementē scdm
speciē suā. Et vidit deus q̄ esset bonū:
et factū est vespere ⁊ mane dies tercius.
Dixitq; aūt deus. Fiant luminaria
in firmamēto celi. ⁊ diuidant diem ac
noctē: ⁊ sint ī signa ⁊ tēpora. ⁊ dies ⁊
annos: ut luceāt in firmamēto celi et
illuminēt terrā. Et factū est ita. Fecitq;
deus duo lumīaria magna: lumīare
maius ut p̄esset diei et lumīare min⁹
ut p̄esset nocti: ⁊ stellas. ⁊ posuit eas ī
firmamēto celi ut lucerēt sup terram: ⁊
p̄essent diei ac nocti: ⁊ diuiderēt lucem
ac tenebras. Et vidit de⁹ q̄ esset bonū:
et factū ē vespere et mane dies quart⁹.
Dixit etiam deus. Producant aque
reptile anime viuentis et volatile sup
terram: sub firmamēto celi. Creauitq;
deus cete grandia. et omnē animā vi-
uētem atq; motabilem quā produxe-
rant aque in species suas: ⁊ omē volatile secundū genus suū. Et vidit deus q̄ esset bonū: benedixitq; ei dicens.
Crescite et multiplicamini. et replete a-
quas maris: auesq; multiplicentur
super terram. Et factū ē vespere ⁊ mane
dies quītus. Dixit quoq; deus. Pro-
ducat terra animā viuentem in gene-
re suo: iumenta ⁊ reptilia. ⁊ bestias terre secundū species suas. Factū ē ita. Et
fecit deus bestias terre iuxta species suas: iumenta ⁊ omne reptile terre in ge-
nere suo. Et vidit deus q̄ esset bonū:
et ait. Faciam⁹ hominē ad ymaginē ⁊
silitudinē nostrā. ⁊ p̄sit piscib; maris.
⁊ volatilib; celi. ⁊ bestijs vniūsq; terre:
omniq; reptili qd mouet ī terra. Et creauit deus hominē ad ymaginē et simi-
litudinē suam: ad ymaginem dei creauit illū: masculū et feminā creauit eos.
Benedixitq; illi deus. et ait. Crescite
et multiplicamini et replete terram. et
subicite eam: ⁊ dominamini piscibus
maris. ⁊ volatilibus celi: ⁊ vniūersis
animātibus que mouentur sup terrā.
Dixitq; deus. Ecce dedi vobis omnē
herbam afferentem semen sup terram.
et vniūsa ligna que habēt ī semetipis
semētē generis sui: ut sint vobis ī escā.
⁊ cūctis aīantibus tre. omniq; volucri
celi ⁊ vniūersis q̄ mouentur in terra. et ī
quibus ē anima viuēs: ut habeāt ad
vescendū. Et factū est ita. Viditq; deus
cuncta que fecerat: et erāt valde bona.

The Book of Genesis, the first book of the Bible's Old Testament, relates the Judeo-Christian version of creation. This page of Genesis was printed in Mainz, Germany, circa 1455 by Johannes Gutenberg, inventor of movable type.

is born, such a recitation is performed by the medicine man, who then touches the child's body to prepare it for life. The same process—including the creation myth—is repeated before the baby drinks water or eats solid food (see also **Osage Creation**).

Blackfoot Creation

The Blackfoot Indians of northern Montana recognize a creator god they call Napi, or Old Man. Old Man creates *ex nihilo* (see also **Creation from Nothing**) and from elements such as clay (see also **Creation from Clay**). His work, as is the case in the myths of many patriarchal cultures, is undermined by a woman (see also **Eve**).

Old Man traveled from place to place creating mountains, valleys, deserts, plants, and animals as he went. He made his way north, where he created the Teton River. After he crossed the river he lay down to rest on his back with his arms extended out from his body and placed stones all around the parts of his body. These stones are still there. Farther north he stumbled over a knoll and landed on his knees. To mark the place he made two great buttes called the Knees, and they are still there too. Farther north he made the Sweet Grass Hills, the prairies, the bighorn, and the antelope.

Taking some clay one day, he made a woman and a child, named them people, and covered them (see also **Creation from Clay**). When later he took the cover away he saw the people had changed. He covered them again several times, and each time he uncovered them they had changed more. Now there were lots of people in addition to the first woman and child. Napi told them to get up and walk, and he introduced himself to them.

Then the woman asked Napi whether the people would live forever, and he answered that he did not know. "I will throw a buffalo chip into the river. If it floats, the people will die, but only for four days. Then they will come back." If the chip sinks, people's lives will end. When he threw in the chip it floated, but the woman was not satisfied. She insisted on throwing a stone into the river. "If it floats," she said, "we must live forever. If it sinks, people will have to die, but they will feel sorry for each other." When the stone sank, Old Man announced that death would end all lives, and the people were sorry for each other.

In the time that followed, Old Man taught the people how to live. He taught them how to hunt, how to use animals as food and clothing, and how to respect the animals. He taught them about fire and cooking and how to gain power from sleep and from the lessons taught by animals in dreams.

Old Man kept traveling north, and the people and animals followed him. One day he came to a steep hill, climbed it, and then, for fun, slid

down it. It is still called Old Man's Sliding Place, and it was near here that the Blackfeet settled.
Source: Hamilton, 25–27.

Blood Indian Creation

The Bloods are Canadian Indians who belong to the Blackfoot Confederacy. Like the Blackfeet, they believe in the creator, Old Man, whom they call Napioa (see also **Blackfoot Creation**). Theirs is an **earth-diver creation** story.

Napioa, who floated about on the first waters on a log, sent the fish, the frog, the lizard, and the turtle to get whatever there was down below. The fish, the frog, and the lizard did not come back, but the turtle did, and he carried some mud with him. He rolled up this mud into a ball, and it grew to become the earth.

Napioa made all of the world except for the white men, and nobody knows where they came from.

After the earth was made it was time for humans. Napioa made a woman, but he made her mouth the wrong way around and had to repair it before making some men and then more women. The men were afraid of the women, but Old Man told them what to do and the couples married.

Finally, Napioa made the buffalo and taught people how to hunt them. Now he lives far away in a southern sea.
Source: Sproul, 244.

Boshongo Creation

The Boshongo are Bantu people of Central Africa. Their creation myth is the type that involves a god creating out of his own bodily fluids—in this case vomit (see also **Creation from Secretion**). In short, this is an example of creation *ex nihilo* (see also **Creation from Nothing**).

In the beginning there was only darkness, water, and the great Bumba. One day, suddenly feeling a pain in his stomach, Bumba vomited up the sun. The sun shone on the water so hard that it began to dry up, leaving land. Then Bumba vomited up the moon and the stars and later various animals, many named after him. There were, for instance, the leopard (Koy Bumba), the crocodile (Ganda Bumba), and the tortoise (Kono Bumba). Men came last, and one, Loko Yima, was white like Bumba.

The animals that had come out of Bumba created other animals. The crocodile made snakes, the heron made birds, and so forth. Three sons of

Bumba also created beings. Only Tsetse (lightning) caused trouble. She was so bad that she had to be confined to the sky, from which place she still sometimes strikes her old home in anger.
Source: Leach, 145–147.

Brahma

Brahma is a version of the Hindu creator. Over the centuries he has become less important than the other two members of the Hindu trinity, Shiva and Vishnu (see also **Indian Creation**).

Brahman

Brahman is sometimes confused with the creator god Brahma, and in a sense this is appropriate as all that is must be suffused with Brahman. Brahman, however, is a neuter concept encompassing and transcending gender, whereas Brahma is only Brahman's masculine form; Brahman is neither male nor female and is everywhere and nowhere. Brahman is transcendent and immanent, the source of all forms. It could be said that all gods are personifications of the Absolute, which is Brahman—the ultimate source of creation, the cosmic power within and behind existence. In the individual, Brahman is Atman or perfect Self. It might be said that Brahman is the Hindu idea of Logos or the Word: "In the beginning was the Word," or in this case Om, the sacred syllable denoting the wholeness of creation.

Buddhist Creation

Buddhism places little emphasis on creation as such, but in its early Indian form, known as Theravada Buddhism, there is a large body of scripture attributed to the Buddha himself. In this writing, called the Pitaka, there is a section called the Digha Nikaya, in which the Buddha speaks of the end of the world and a new creation (see also **Buriat Creation**). His vision contains several familiar elements of creation myths from elsewhere, beginning with the dominance of the primeval waters. There is no creator as such, and creation is *ex nihilo* (see also **Creation from Nothing**).

In time our world will come to an end, but also in time the world will evolve again. Then everything will be covered in water and darkness. For a long time there will be no sun, moon, stars, or seasons, and there will be no

A relief carved in the second century, shows the birth of Buddha, Siddhartha Gautama, whose life and writings are the basis of Buddhism.

creatures, no humans. After a still longer time, earth will form on the waters, as skin forms on cooling hot milk. Then some greedy being from a former birth will forsake the heavenly radiance of the Buddha soul-life for the life of the body and will take pleasure in the earth's sweetness. Then others will follow, and they will gradually become more body than radiant Buddha soul. As their light fades, the sun, moon, and stars will appear. Gradually humans will develop sexual characteristics, which will be followed by passion and then by selfishness and other evils. In time the new world will come to its end.

Source: Embree, 133–138.

Bulu Creation

The *ex nihilo* creation myth of the Bulu people of Cameroun is deeply affected by a negative self-image that is clearly the result of colonialism (see also **Creation from Nothing**).

In the beginning was Membe'e, he who holds up the world. His son Zambe was sent to create Man, Chimpanzee, Gorilla, and Elephant, each of whom he named after himself. One of the men he created was black and another was white. Zambe gave the new Zambes many fine things, such as water, gardening tools, and especially fire and the book.

The new beings stirred the fire. When smoke got in the white man's eyes, he went away with the book. Chimpanzee left the fire and the other gifts and went into the forest to eat the fruit there. Gorilla soon followed his lead, and Elephant just stood around not thinking about much of anything. As for the black man, he continued to stir the fire, but he didn't bother about the book.

When the creator came for a visit he called his creatures together and asked what they had done with the things he had given them. When Chimpanzee and Gorilla said what they had done, Zambe condemned them to having hairy bodies and big teeth and to live forever in the forest eating fruit. Elephant was sent off in much the same manner.

Zambe now asked the black man where his book was. The black man replied that he had not had time to read it because he was tending the fire. "Well," said Zambe, "that's what you will continue to do; you will have to spend your life working hard for others because you do not have book knowledge."

Zambe turned to the white man and asked him what he had done with the gifts. "I have only read the book," he said. "And that you shall continue to do," answered the god. "You will know lots of things, but you will need the black man to take care of you because you will know nothing about keeping warm and growing food."

So it is that the animals live in the forest, white men sit about reading a lot, and black men work hard but always have a good fire going.

Sources: Leach, 140–142; Sproul, 45–46.

Buriat Creation

This is apparently the earliest version of a creation story that is told in different versions by various peoples in Siberia (see also **Altaic Creation; Samoyed Creation**). It is an earth-diver creation (see also **Earth-Diver Creation**) as well as a "How-the-Leopard-Got-Its-Spots" origin story (see also **Origin Stories**), and it also contains the creation from clay motif.

In the beginning there were only the waters and the god Sombov until the god saw the water bird Anghir. Sombov ordered the bird to dive into the waters and bring back some earth. The bird returned carrying both black earth and red clay. Out of the first, Sombov molded the Earth; out of the clay he made two wool-covered beings—man and woman—but he decided not to give them life until he had obtained souls for them (see also **Adam**). While he went off to heaven to get the souls, he left a dog—at this point in creation still without fur—to watch over his unfinished work. The dog, shivering in the cold, did as he was told, but Shiktur, the devil, came by and promised the dog a fur coat if he would let him see the new humans. The dog gave in, and Shiktur fouled the new creations by spitting on them. When Sombov returned, he was, of course, furious. For its disobedience, the dog was condemned to a life of shivering in spite of its coat. Wherever the devil's spit had touched the humans the wool had to be removed from their flesh, leaving them—especially the woman—naked in all but certain parts of their bodies. At least the humans were given life and souls.

The Mongol-related Buriats are strongly influenced by Buddhism. In this earth-diver version of the god-devil creation (see also **Earth-Diver Creation**), the creator, Burkhan, is the Buddha. The vision of the creation contained in it is one that has evil within it from the beginning; it is a cooperative venture of god and the devil.

Burkhan came from heaven to create the earth and was met by Sholmo, the devil. Sholmo offered to dive under the primordial waters to find the material Burkhan would need for his work. When the devil returned with the material, Burkhan made the earth by scattering dirt and stones on the

sea. The devil asked for a bit of the earth as a place where he could plant his staff. From the hole he made with the staff, all the evil creatures of the world emerged, especially snakes.
Sources: Leach, 200–201; Sproul, 219.

Bushmen Creation

The Bushmen, who live in the Kalahari Desert of southwest Africa, believe in an *ex nihilo* creator god who takes the form of the praying mantis and whose name sounds like "kaggen" (sometimes written as Cag-n), with the typical clicking sound before it. In fact, the term means praying mantis, and that insect is sacred to the Bushmen.

Mantis, as we can call him, was the creator of almost everything, and in the old days he lived here with humankind. It was the foolishness of humans that drove him away in disgust and left so many of us hungry.

Many stories are told of the creator in those days long ago. We hear of his wife Coti and two sons who taught the people how to find food in the earth, and of a daughter who married a snake. It is said that Mantis could become any animal he wanted, but most of all he liked becoming an eland bull. The elands are still his favorites, and only they know where he is. People also say that Mantis created the moon by throwing his shoe into the night sky.
Source: Leach, 152–155.

California Indian Creation

There are several examples of creation myths from individual California tribes (see also **Maidu Creation; Pomo Creation; Yokut Creation**), but the many tribes of California were so decimated by early European settlers that it is also helpful to see them collectively as a group possessing mythological fragments that, when gathered together, form a trickster-creator Coyote cycle (see also **Coyote; Trickster**).

A Miwok myth, for example, tells of Coyote creating the earth by shaking his blanket over the waters. Coyote is the source of language in a Wappo myth in which he presents the people with a bag of words from the Great Spirit. In some myths Coyote creates the people themselves, from sticks in a sweathouse, for instance.

Source: Bierhorst, 116.

Catastrophic Creation Theory

See Big Bang Theory.

Celtic Creation

Not much is known about the myths of the early Celts who descended into Europe and the British Isles and Ireland in the latter part of the pre-Christian era. A creation myth has been inferred from early Roman writings and archeological finds. It seems to indicate a connection between the Celts and the Aryans, or Indo-Europeans, who invaded the Mediterranean world many centuries earlier.

The inferred myth depicts Heaven and Earth as the original parents, reminding us not only of Geb and Nut in Egypt (see also **Egyptian Creation**) but Gaia and Uranos in Greece (see also **Greek Creation**). The first pair is so close that there is little room for creation between them. One of the evil children of Heaven and Earth separates the pair by castrating the father, from whose skull the gods' children made the sky and from whose blood they made the sea (see also **World Parent Creation**). The son who had castrated the father became god of the underworld; the good children became the gods of the sky and earth.
Source: Sproul, 172–173.

Central Asian Creation

A well-known Central Asian **earth-diver creation** myth has clear connections to the many earth-diver myths of the Native Americans and to the Mongolian-related tribes of Siberia (see also **Buriat Creation; Huron Creation**).

The creator, Otshirvani, and his helper, Chagan-Shukuty, came down from above and noticed a frog diving into the water. Chagan-Shukuty reached down for it and turned it onto its back so Otshirvani could sit on its stomach. Otshirvani ordered his assistant to dive to the bottom of the waters and bring back whatever he found there. After several attempts, Chagan-Shukuty brought back some earth, and Otshirvani ordered him to sprinkle it on the frog's stomach. Now the frog sank a bit under the weight, leaving only the earth visible, and the two gods rested on the new earth. While they were sleeping, the devil came by and decided to destroy the gods and their new earth. He picked up the sleeping creators and ran with them towards what he thought would be the waters, but the farther he ran the more the earth grew. Finally, he dropped the sacred beings, who woke up. Otshirvani explained to his companion what had happened and praised the new earth for saving them.
Source: Long, 205–206.

Ceram Creation

The Ceram people live in the Molucca Islands of Indonesia. Theirs is an agricultural society dependent upon successful harvesting, a cutting of the plants. Appropriately, their creation myth is an example of **creation by emergence** from the earth and of sacrifice, dismemberment, and planting (see also **Creation from Dismemberment of Primordial Being**).

The nine original families emerged from bunches of bananas and then came down from Mount Nunusaka to the place now called Nine Dance Grounds in West Ceram. One man, Ameta, was much darker than the others, and he was very much a loner. He went hunting one day and killed a wild pig with a coconut caught on its tusk. No one had ever seen coconuts or coconut trees at that time, so Ameta took it home and wrapped it for safekeeping in a cloth designed with a snake figure. That night a man came to him in his dreams and instructed him to bury the nut. This Ameta did in the morning, and within days it was a fine, tall palm bearing coconut blossoms. Ameta climbed the tree to harvest some fruit but cut his finger. When he returned to the tree after fixing his cut, he found that his blood had mixed with the tree's sap to form a face, and in a few days he found a little girl there. The dream man appeared to Ameta in the night and told him to wrap the girl in his snake cloth and bring her home. This Ameta did, and he named the girl Hainuwele. In a few days Hainuwele was grown, and amazingly, she defecated things like dishes and bells, which her father sold (see also **Creation by Secretion**).

It then came time for the nine families to perform the nine nights of the Maro dance at Nine Dance Grounds. As was customary, the women of the families sat in the center of the dance grounds handing out betel nut to the men, who danced around them in a spiral. Hainuwele was at the very center.

On the first night she handed out betel nut, but on the second she gave the dancers coral instead, and on the third night she gave out fine pottery. In fact, she gave out more and more valuable objects each night. The people became jealous of her obvious wealth and decided to kill her. On the ninth night, having dug a deep hole at the center of the dance place, they surrounded her during the dance and edged her into the hole and covered her with earth.

Ameta missed his daughter and, guessing that something had happened to her, used his oracular skills to discover that she had been killed during the Maro dance in Nine Dance Grounds. He took nine pieces of palm leaf to the grounds and stuck them into the earth. The ninth one he placed at the very center of the grounds, and sure enough, when he pulled it out he found bits of his daughter's blood and hair. He dug up the body, cut it into many pieces, and buried all but the arms in the dance grounds. Immediately there grew the plants that are the staples of the Ceram people to this day.

Ameta took Hainuwele's arms to the goddess Satene, who then went to Nine Dance Grounds, built a huge gate there and stood behind it holding out the maiden's arms. She called the nine families and announced to them that in revenge for their killing of Hainuwele she would leave them, but that everyone would first have to try to pass through the gate to her. Those who succeeded would remain people, those who did not would become animals and spirits. So it was that animals and spirits came into being. Satene then traveled to the Mountain of the Dead, where anyone who follows her must die.

The Ceram people tell of another sacred maiden like Hainuwele. Her name was Rabia, and she was taken away by the sun god, Tuwale. It was she who instituted the tradition of the Death Feast and who became the moon.
Source: Long, 224–229.

Chaos

Chaos is the Greek word for the dark void of pre-creation (see also **Creation from Chaos; Greek Creation**).

Chaos to Cosmos

This is the essence of all creation, in which order (cosmos) is wrenched by some force or other from chaos (the void) (see also **Creation from Chaos; Creation from Nothing**).

Cherokee Creation

The Cherokee Indians or Tsalagi (the People) were originally from the southeastern section of what is now the United States and later were forced to move west to Oklahoma. They tell several creation stories, usually dominated by a female sun. The tales are, for the most part, **earth-diver creation** myths, but there are elements of the emergence myth as well (see also **Creation by Emergence**). They also contain the popular motif of the sun-catcher. The myths are always told at night and in winter so the fire of life might be rekindled in the listener.

There was a time long ago when everything was covered by water. Anything that was alive then lived in Galunlati, the vault above the sky, where it was so crowded that it was hard even to move. Desperate for more space, the animals sent Water Beetle out to explore. He dove to the bottom of the

Twentieth-century Cherokee artist and storyteller Jimalee Burton painted a version of her people's creation. The first people, who had lived as stars in the heavens, arrive on an earth covered by water. High winds blow the water into a mist, leaving dry land.

waters and came back to the surface with a bit of mud. The mud spread out and became the earth-island, which the Great Spirit fastened to the rock sky with four rawhide cords stretching from the four sacred mountains of the four sacred directions.

The earth was still muddy, though—too soft to hold the weight of the creatures. Buzzard was sent to find a dry spot, and eventually he came to one that was at least beginning to dry. This was the place that would become the Cherokee country. That country has many mountains and valleys, created by the furious movements of Grandfather Buzzard's wings.

When the new country was dry and hard enough, the animals descended from the vault above the rainbow, but they were bothered by the darkness in their land. They decided to pull Sister Sun down from behind the rainbow. They did this and then assigned her a regular path to follow.

It is said, too, that the Great Spirit made all the plants to go with the animals, and that he asked them to stay awake for seven days. He asked the animals to stay awake, too. Most of the animals fell asleep before the eighth day, but the owl stayed awake and was given the power of night sight. The few plants that did not sleep—the pine, the holly, the laurel, and a few others—were allowed to keep their hair all through the winter. The other plants shed each year.

The Great Spirit also made a man and a woman. The man pushed a fish against the woman and made her pregnant, and she gave birth to a child every seven days until the Great Spirit regulated things so she could only give birth once a year.

Many people say it was Star Woman who was the primal cause. One story says she was in her father's garden, that is in Galunlati, when she heard drumming under a tree and dug a hole to see what was going on. Star Woman fell through the hole and spun toward the earth. At that time the earth was under the primeval flood, and earth creatures lacked the spark of deep consciousness or understanding. They did have feelings, however.

The father watched his daughter fall and called on the winds to get the earth creatures to help her. Turtle suggested that his back become a landing place for her, so the animals dove into the depths to find something soft to place on Turtle's back. Only Water Spider—some say Muskrat—succeeded. She brought up a bit of earth and placed it with her last bit of strength on Turtle's back; then she sank to her death.

Now the earth on Turtle's back grew, and Buzzard made mountains and other beautiful places by stirring the earth up through the flapping of his great wings. All was ready for Star Woman, who landed on Turtle's back and immediately produced corn, beans, other plants, and rivers from her body (see also **Animism**). Most of all, she brought the spark of consciousness,

symbolized by the Cherokees' sacred fire, which is always kept alive for the ceremonies.
Sources: Erdoes, 105–107; Ywahoo, 29–30.

Cheyenne Creation

The Cheyenne are American Plains people whose creation story is an epic of constant migration, reflecting the history of the tribe's reactions to natural disasters and threats from other humans.

In the beginning the Great Power made a beautiful world of earth, waters, and sky. Up north he made the most perfect place—a paradise where it was always warm and where there was always plenty of good food and water. In that land, all the animals and people could understand each other and were friends. There was no need of shelter or clothes.

The Great Power had created white people, hairy people, and red people. The hairy people were shy; they lived in caves, and eventually they disappeared. The white people were clever and tricky; the red people were close to the Great Power, who eventually told them to band together and travel to the more barren south. This they did, and, since the Great Power taught them how to hunt and to make clothes, they thrived until the Great Power warned them of a coming flood (see also **The Flood and Flood Hero**). To avoid the flood they moved back to the old north country, but they could no longer talk to the animals there. Still, they did well and improved their hunting skills.

Later the red people moved south again, but another flood came and they were scattered into the many tribes that exist today. They returned to the north, but their old land was desolate, so they returned south. After many years there, an earthquake and volcanoes nearly destroyed the tribes. After things calmed down, the Great Medicine Man in the sky gave them buffalo and corn, and some of the people established themselves in what is now the Cheyenne country.
Source: Erdoes, 111–114.

Chinese Creation

There are several versions of the Chinese creation myth, the primary sources of which are popular legend and a third-century literary text entitled *San-wu li-chi.* In this myth we find several familiar elements, among them the creation from a primal cosmic egg, the separation of heaven and

Phan Ku was the first man on earth according to Chinese tradtion. He is usually represented as a rough and hairy giant who separated heaven and earth and the other opposites—yin and yang—such as light and darkness, hot and cold, male and female.

earth, and the ordering principle, yin-yang—the everywhere and nowhere, the Logos, the Word, Brahman-Om. The world left by the creator is an animistic one made up of his being, the embodiment of yin and yang (see also **Animism; Creation from a Cosmic Egg; Creation from Division of Primordial Unity**).

In the beginning was a huge egg containing chaos, a mixture of yin-yang—female-male, passive-active, cold-heat, dark-light, and wet-dry. Within this yin-yang was Phan Ku, that which was not yet anything but which broke forth from the egg as the giant who separated chaos into the many opposites, including earth and sky. Each day for 18,000 years Phan Ku grew ten feet between the sky, which was raised ten feet, and the earth, which grew by ten feet. So it is that heaven and earth are now separated by 90,000 li, or 30,000 miles.

Phan Ku was covered with hair; horns sprang from his head and tusks from his mouth. With a great chisel and a huge mallet, he carved out the mountains, valleys, rivers, and oceans. During his 18,000 years he also made the sun, moon, and stars. He taught the people what they know. All was suffused by the great primal principles of the original chaos, yin and yang.

When Phan Ku finally died, his skull became the top of the sky, his breath the wind, his voice thunder, his legs and arms the four directions, his flesh the soil, his blood the rivers, and so forth. The people say that the fleas in his hair became human beings. Everything that is is Phan Ku, and everything that Phan Ku is is yin-yang. With Phan Ku's death a vacuum was created, and within this vacuum pain and sin were able to flourish.

A more philosophical creation from about 200 B.C.E. also begins with chaos (see also **Creation from Chaos**).

In the beginning was chaos, from which light became the sky and darkness formed the earth. Yang and yin are contained in light and darkness, and everything is made of these principles.

When yang and yin became one and the five elements were separated, humankind was born. As the first man watched the patterns of the sun, moon and stars, a gold being came down and stood before him. The Gold One taught the man—now named the Old Yellow One—how to stay alive and how to read the sky.

He explained the beginning of things to the Old Yellow One. He explained how the life force that flows though us was created by earth and heaven, how the relative power of yin and yang at any given time results in heat or cold. He explained how the sun and moon trade light, how this act causes the passing of time and how it creates the four directions and the midpoint. He said that the sky and earth together produced man and that the yang principle gave and the yin principle received.

The Gold One told Old Yellow One about a great stone in earth's center and about the poles that support the earth. He told how the waters of earth surround it the way the flesh of fruit surrounds the seed. It is in this way that all things correspond—the fruit, the egg, the earth, the body—all things.

The great Chinese philosopher and founder of Taoism, Lao-Tzu, presents hints of cosmological thinking in the sixth-century B.C.E. work, the Tao te Ching. For the Taoist, creation is the Way, the Tao itself. Being for the Taoist comes from Non-Being.

This philosophy is illustrated in the following segment of the Tao te Ching.

> There is a thing confusedly formed,
> Born before heaven and earth.
> Silent and void
> It stands alone and does not change,
> Goes round and does not weary.
> It is capable of being the mother of the world.
> I know not its name
> So I style it "the way."
> I give it the makeshift name of "the great."
> Being great, it is further described as receding,
> Receding, it is described as far away,
> Being far away, it is described as turning back.
> Hence the way is great; heaven is great; earth is great; and the king is also great.
> Within the realm there are four things that are great, and the king counts as one.
> Man models himself on earth,
> Earth on heaven,
> Heaven on the way,
> And the way on that which is naturally so.
>
> Reprinted from D. C. Lau, trans., Tao te Ching, Baltimore, MD: Penguin, 1963.

Sources: Calum, 237–239; Kramer (A), 332; Long, 126–128; Sproul, 199–200.

Christian Creation

A creation story appears at the beginning of the Gospel of John in the New Testament, the Christian addition to the Bible. Not intended as

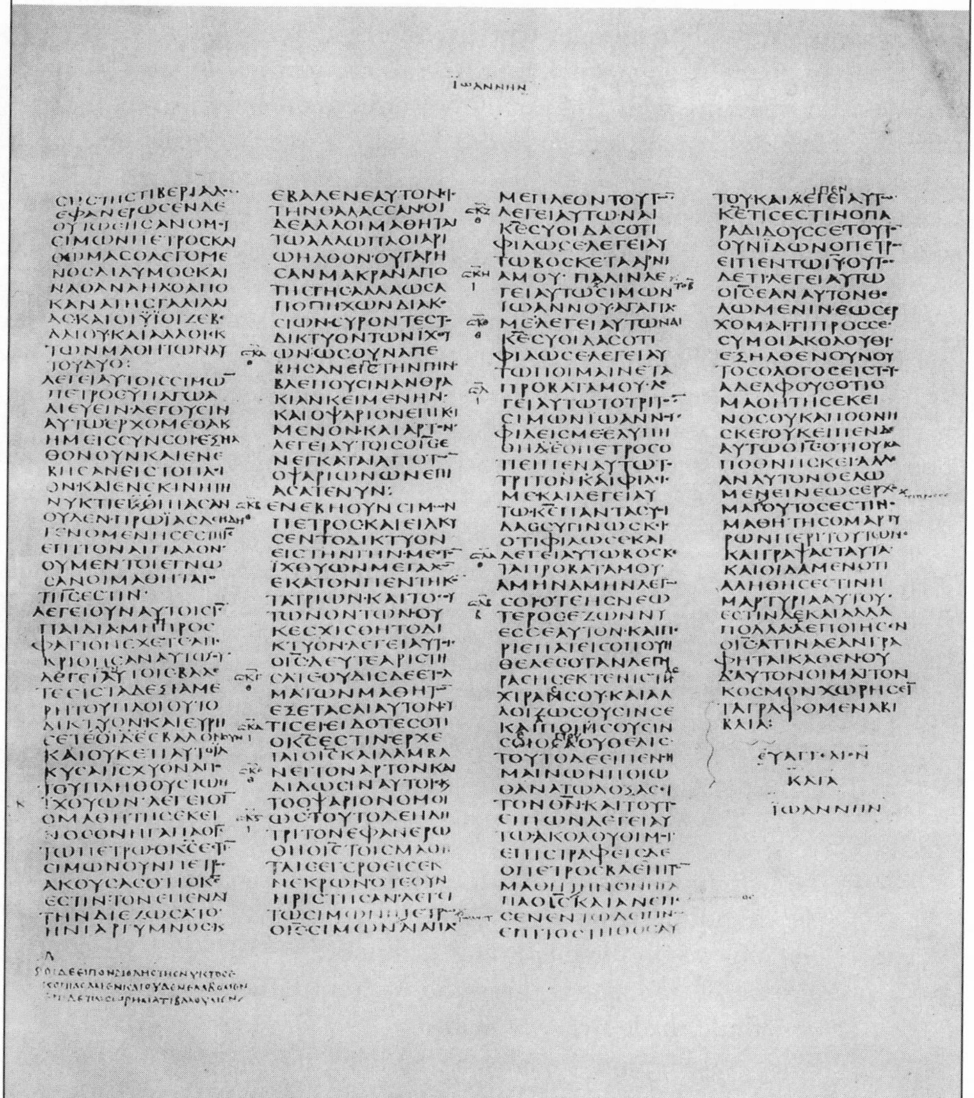

ΙΩΑΝΝΗΝ

[illegible]

ΕΥΑΓΓΕΛΙΟΝ
ΚΑΤΑ
ΙΩΑΝΝΗΝ

The Bible's New Testament contains the gospel according to Saint John, which presents a new way of seeing the creation of the Old Testament's Book of Genesis. This is a page of the Codex Sinaiticus written in Greek late in the fourth century.

an alternative or substitution for the Genesis account, John's gospel was seen by early Christians as a more "spiritual" document than the "factual" gospels of Matthew, Mark, and Luke. Some say John was the man often spoken of as the favorite disciple of Jesus. Some say he wrote the Book of Revelation, too. Others say he was a church leader who lived in Ephesus in about 100 C.E.

In the prologue to his gospel, John reveals the identity of Jesus as Logos (the Word). Logos in Greek philosophy was cosmic order or reason, the creative force behind cosmos (universal order), which was originally developed out of chaos (the void)—or, *ex nihilo* (see also **Creation from Chaos; Creation from Nothing**). For John, then, Jesus is the human form of the everexisting divine Logos that created the universe: "In the beginning was the Word."

The Word existed before anything else, and the Word and God were one and the same. All creation came about through the Word. All life is alive with his life, his being. His life is the light that shines in the human being; that is, he is the source of human consciousness, the human soul. The Word sent John the Baptist to the world to prepare it for his coming. Then the Word was "made flesh" as Jesus.

Later Christians would sometimes refer to Jesus as the New Adam, he who represented the possibility of a new beginning, a new creation based on Christian belief and principles. It is said he overcame death through his death and, as one prayer has it, "made the whole creation new."

The following is the beginning of John's gospel as translated in the King James Version.

> The Gospel According to Saint John
>
> Chapter 1
> In the beginning was the Word, and the Word was with God, and the Word was God.
> [2] The same was in the beginning with God.
> [3] All things were made by him; and without him was not any thing made that was made.
> [4] In him was life; and the life was the light of men.
> [5] And the light shineth in darkness; and the darkness comprehended it not.
> [6] There was a man sent from God, whose name *was* John.
> [7] The same came for a witness, to bear witness of the Light, that all *men* through him might believe.
> [8] He was not that Light, but *was sent* to bear witness of that Light.

[9] *That* was the true Light, which lighteth every man that cometh into the world.
[10] He was in the world, and the world was made by him, and the world knew him not.
[11] He came unto his own, and his own received him not.
[12] But as many as received him, to them gave he power to become the sons of God, *even* to them that believe on his name:
[13] Which were born, not of blood, nor of the will of the flesh, nor of the will of man, but of God.
[14] And the Word was made flesh, and dwelt among us, (and we beheld his glory, the glory as of the only begotten of the Father,) full of grace and truth.

Source: Bible, John I.

Chukchee Creation

This Eskimo myth from northeastern Siberia is an example of creation by defecation, perhaps a form of **creation from nothing**. It features Raven, a figure popular not only in Siberia but among the tribes of North America, especially the Eskimos (see also **Creation by Secretion**). Creation by defecation, urination, and unbridled eroticism is typical of the creative but amoral **trickster** archetype of which Raven is almost always an example. Scholars of a Freudian orientation have tended to see male anal creation myths as examples of the male's envy of the woman's ability to procreate. Freud might have called it "vulva envy."

In the beginning Raven, the self-created, lived with his wife in a tiny space. Bored with her existence, the wife asked Raven to create the earth. "But I can't," he said. "Well, then," said the wife, "I shall create at least something." She lay down to sleep, with Raven watching over her. As she slept, the wife seemed to lose her feathers and then to grow very fat, and then without even waking up she released twins from her body. Like the mother, now, they had no feathers. Raven was horrified, and when the twins noticed him they woke their mother and asked, "What's that?" She said "It's father." The children laughed at the father because of his strange harsh voice and his feathers, but the mother told them to stop, and they did.

Raven felt he must create something since his wife had created humans so easily. First he flew to the Benevolent Ones—Dawn, Sunset, Evening and the others—for advice, but they had none to give. So he flew on to where some strange beings sat. They were to be the seeds, they said, of the new people, but they needed an earth. Could Raven create one? Raven said he would try, and he and one of the man-seeds flew off together. As he flew Raven defecated and urinated, and his droppings became the mountains, valleys, rivers, oceans, and lakes. His excrement became the world we live in. The man-seed with him asked Raven what the people would eat, and Raven made plants and animals.

Eventually there were many men from the original seed, but there were no women until a little **spider woman** appeared and made women. The men did not understand about women, so Raven with great pleasure demonstrated copulation with the women; later, also with pleasure, the men followed his example.

Another Chuckchee story says that in the beginning there were two beings, Raven and Creator, that Raven told Creator to make a man, and Creator did as he was told. The man was animal-like—hairy and four-legged, with great claws and teeth. He could catch any animal he hunted, and he ate everything raw. Creator feared that he would destroy all of living creation, so Raven suggested that they slow man down and make him less dangerous by shortening his arms and making him walk upright. They also substituted clothes for hair. So man could eat regularly, Creator made reindeer out of various plants and made the people herders. So the people could get around, he created dogs out of wood, and he gave the people the characteristic Chuckchee walking stick for support. It happened that one Chuckchee family lost a dog and Raven found it. That's why Raven has a dog now.
Sources: Leach 198–199; Weigle, 228–231.

Collision Creation Theory

The collision creation theory is one of the creation "myths" of science. It was developed by British scientist Harold Jeffreys from an earlier theory. The theory postulates a collision between the sun and a large out-of-orbit star. The crash caused bits of the sun to break off and eventually form the planets (see also **Creation from Chaos; Creation in Science**).
Source: Leach, 19.

Continuous Creation Theory

The continuous creation theory gained a great deal of support among British cosmologists in the mid-twentieth century. According to this

theory, hydrogen atoms are created each year in distant space. The atoms are attracted to each other by gravitation until they form a universe of stars and galaxies. When gravitation fails, the atoms break away from each other and disappear into space. More hydrogen atoms are formed, and the whole process begins again. The theory might be called the breathing universe theory (see also **Creation from Nothing; Creation in Science**).
Source: Leach, 19–20.

Cosmic Creation Myth

This is a new-age, pseudo-scientific primal egg myth based on the thoughts and writings of philosopher Thomas Berry (see also **Creation from a Cosmic Egg**). It is oriented toward the big bang theory and is contained in a book called *The Universe Is a Green Dragon,* by physicist Brian Swimme. The book is essentially a Socratic dialogue between Swimme as student and Berry as teacher. It explores the idea of "cosmic allurement," the cosmic "love" that binds all nature together.

> ...the universe is a green dragon. Green, because the whole universe is alive, an embryogenesis beginning with the cosmic egg of the primeval fireball and culminating in the present emergent reality. And a dragon, too, nothing less. Dragons are mystical, powerful, emerging out of mystery, disappearing in mystery, fierce, benign, known to teach humans the deepest reaches of wisdom. And dragons are filled with fire. Though there are no dragons, we are dragon fire. We are the creative, scintillating, searing, healing flame of the awesome and enchanting universe.
>
> Reprinted from Brian Swimme, *The Universe Is a Green Dragon,* Santa Fe: Bear & Co., 1984.

Source: Swimme.

Cosmogony

The word *cosmogony* comes from the Greek *kosmogonia,* meaning creation of the world (from *-gonos,* derived from the verb meaning produce, and *kosmos,* meaning cosmos, universe or world.)

A cosmogony is a creation story, a culture's account of its cosmic origins. Each cosmogony gives spiritual or cosmic significance to the given culture's surroundings and activities. It establishes the culture's importance and establishes its environment as the center of the world. It is true, of course, that a given culture's cosmogony is seen by others as its creation myth.

Cosmology

The word *cosmology* comes from the Greek *kosmologia,* meaning study of the cosmos, or universe. Cosmology is that branch of science or philosophy that concerns itself with the study of the universe as a system. A cosmogony is, in this sense, an aspect of cosmology.

Coyote

Coyote—sometimes called Old Man—is one of the prevalent forms of the **trickster** figure among Native Americans. He is often the assisting agent by whom the supreme deity creates the world. Sometimes he works mischievously in creation, playing the role of a devil-like or at least amoral figure who uses his unbridled sexuality and excretory powers to bring about changes that become, in effect, aspects of creation (see also **Crow Indian Creation**). Raven is another popular trickster mask, as is Inktomi (see also **Chuckchee Creation; Inktomi; Sioux Creation**).

Creation

The word *creation* can refer to the world or universe, that which the creator—any supreme god—has created, or to the act of creation itself. We can speak, therefore, of "all the animals in creation" or of the "six days of creation" (see also **Cosmogony**).

Creation by Deus Faber

Deus faber, or *Dea faber,* is the creator in his/her form as craftsman or artist. Jungian analyst Marie-Louise von Franz, in her *Patterns of Creativity Mirrored in Creation Myths* (1972), suggests that in *Deus faber* creations, God creates the world "on the analogy of some skill or craft." We are

Judeo-Christian belief characterizes God as architect of the universe. Here God, in an illustration from the Bible Moralisee compiled in Paris circa 1250, uses dividers to measure the universe.

reminded of the Book of Job (38:4–5) when Yahweh refers to his having "laid the foundation of the earth" and of having "determined its measurements." A popular depiction of the Judeo-Christian God is an architect measuring out the universe with a compass (see also **Hebrew Creation; Huron Creation; Spider Woman; Yuki Creation**).

Creation by Emergence

Emergence is a basic concept in creation myths that takes various specific forms, especially among the Indians of the American Southwest and Mexico. This type of myth describes the emergence of a people into this world by way of one or more underworlds. The underworld in such creations can be seen as a world womb, a place in the Earth Mother where humans, plants, and animals are conceived and gradually mature from a seedlike state in darkness until they are ready to be born through a sacred opening. In the underworld, the people, especially, undergo a process of development to prepare them for a new life under the sun. Sometimes the underworld "people" are still animals, from which they later take clan names or around which they later develop totem traditions. They are taught by some agency of the supreme forces in nature. The world womb aspect of the myth suggests an earlier time when the earth itself was sacred, the source of all possibilities, when goddess as earth reigned supreme. In this connection, it is of interest to note the presence in many of the emergence myths of a creative female midwife such as Spider Woman or Thinking Woman. The male role in the emergence is slight or sometimes nonexistent (see also **Acoma Creation; Hopi Creation; Navajo Creation**).

Creation by Sacrifice

This type of creation almost always involves the sacrifice of a god. It is closely related to the dismemberment motif (see also **Creation from Dismemberment of Primordial Being**), since many of what might be called divine scapegoats are dismembered in order to begin new creations. For instance, Jesus, the Egyptian god Osiris, the Phrygian god Attis, and the Greek god Dionysos are all actually or symbolically dismembered as part of a process of agricultural or spiritual renewal (see also **Ceram Creation; Icelandic Creation; Indian Creation; Rig Veda; Persian (Iranian) Creation**).

Creation by Secretion

A popular offshoot of the *ex nihilo* type of creation is one in which the god—nearly always male—creates from his own secretions: spit, vomit, semen (via masturbation), sweat, urine, or feces. This type of myth assumes the existence before anything else of a solitary god, a personified male version of the potential for creativity that in other systems is the maternal underworld, the maternal cosmic egg, or the maternal primal waters (see also **Bantu Creation; Boshongo Creation; Chuckchee Creation; Egyptian Creation**).

Creation by Thought

In some *ex nihilo* myths, creation is a projection of the creator's thoughts. This idea is hinted at in the aboriginal Dreaming creations (see also **The Dreaming**). Creation by thought presupposes a powerful supreme god who existed before existence itself. It also suggests a mystical sense of the world as contained within the mind of God, the world as a thought that could be forgotten (see also **Buddhist Creation; Laguna Creation; Navajo Creation; Omaha Creation; Tuamotuan Creation; Winnebago Creation; Zuni Creation**).

Creation by Word

In this version of creation *ex nihilo*, the supreme being speaks the Word, making the age-old connection between Logos or cosmic order (the Word) and the ordering principle, which is language (words). So it is that the Hebrew creator, Yahweh, instructs Adam and Eve—made "in his image"—to be creative by "naming" the other creatures (see also **Christian Creation; Hebrew Creation; Mayan Creation; Navajo Creation**).

Creation from a Cosmic Egg

In many creation myths the great pre-creation void was in the form of an egg. The mythmakers inevitably saw an analogy between creation and a birth process they could easily witness in everyday life. The cosmic egg, however, unlike those of reptiles and birds, was often depicted as silver or gold, like the sun and moon (see also **Chinese Creation; Egyptian Creation; Finnish Creation; Indian Creation; Japanese Creation;**

Mande Creation; Orphic Creation; Pelasgian Creation; Polynesian Creation; Tahitian Creation).

Creation from Ancestors

In many cultures creation was by tribal ancestors. This is particularly true of the Australian aborigines, whose first ancestors "dreamed" their particular "worlds" into existence (see also **Arandan Creation; Djanggawul Creation; The Dreaming; Ngurunderi Creation**).

Creation from Chaos

Chaos is the Greek word for the primal void. In creation myths it is the indeterminate, undifferentiated no-thing-ness before some power or force gives it form and reality and thus turns it into cosmos. Some have included in the concept of chaos the idea that the material of creation was always there along with the potential for creation itself (see also **Babylonian Creation; Chaos; Chaos to Cosmos; Chinese Creation; Egyptian Creation; Germanic Creation; Greek Creation; Japanese Creation; Mixtec Creation**).

Creation from Clay

The creation by clay theme would seem to be among the most ancient, reaching back to those pre-patriarchal times when female creative powers dominated the world. Earth, from which clay comes, is traditionally associated with the world mother. Creation from clay alone would seem to suggest the old Mother Power of the ancient goddess cultures. Later clay myths, in which the male plays a significant role as one who breathes life into the clay or fertilizes it in some way, suggest a movement toward the patriarchal vision (see also **Blackfoot Creation; Dyak Creation; Egyptian Creation; Hebrew Creation; Polynesian Creation; Sumerian Creation; Yoruba Creation**).

Creation from Dismemberment of Primordial Being

These myths describe the cosmos as having been created from the cutting up of a monster (see also **Babylonian Creation; Indian Creation; Rig Veda; Norse Creation; Prose (Elder) Edda**).

Creation from Division of Primordial Unity

In this type of creation, the cosmos is created from the breaking of a cosmic egg, from the division of chaos, or, most popularly, from the separation of original or world parents. In a sense, chaos as earth or the primal waters can be seen as the commingled primal parents, who must be separated for creation to take place, creation being a long process of differentiation. Usually, but not always, the sky is the world father and the earth the world mother. Together they are a mass of no-thing, that is, chaos; separated, they become the world. In many of these myths it is a newer god who performs the separation of the parents (see also **Creation from a Cosmic Egg; Creation from Chaos; Egyptian Creation; Separation of Heaven and Earth; World Parent Creation**).

Creation from Nothing

Sometimes called creation *ex nihilo* or *de novo*, this type of creation is particularly popular in monotheistic religions, but it exists elsewhere as well and is sometimes difficult to differentiate from creation from the primal void or chaos when those terms refer to essential nonexistence. In the *ex nihilo* myth, the god figure creates most often by thinking, speaking, or breathing. Other methods involve bodily secretions, illustrated in Egyptian, Greek, Hebrew, Indian, Mayan, Maori, Tuamotuan, Uitoto, and Zuni creations (see also **Creation by Secretion; Creation from Chaos; The Dreaming**).

Creation in Science

It has been pointed out by cosmologist Philip Freund and others that the many creation theories of modern science are marked by the myth of a "beginningless beginning," that is, scientists, like earlier mythmakers, postulate a pre-creation universe in which something existed as a basis for the form eventually taken by the cosmos. This means that science has not been any more successful than religion at confronting the problem of ultimate origin. Science has, however, given us many hypotheses of creation from undifferentiated equilibrium or the void. These hypotheses, many of which have had their day only to be discredited, are narratives meant to explain a mysterious cosmic phenomenon. As such, they can loosely be called myths (see also **Big Bang Theory; Collision Creation Theory; Continuous**

Creation Theory; Cosmic Creation Myth; Eruption Creation Theory; Gaia Principle; Nebular Creation Theory; Tidal Creation Theory).

Creation Myths as Explanation

Strictly speaking, some myths are more origin myths than cosmogonies; that is, their primary purpose seems to be etiological in some specific sense. Such myths tell how the leopard got its spots, how the dog got its fur, or how a particular spring or other sacred place came into being (see also **Altaic Creation**). There are few creation myths, however, that are not in part explanatory (see also **Etiology**).

Creation Myths in Curing Ceremonies

Creation myths are often an important element in curing ceremonies, for the obvious reason that the result of curing is to allow the formerly sick individual to begin again. To ensure a new beginning, the creation myth is recited or in some way represented. We see this in many Native American curing ceremonies. In ceremonies such as the Blessing Way and the Enemy Way, the Navajo patient, for example, is placed in a sand painting recalling the creation, in order that he/she might be cosmically reoriented. An important part of this curing is the shaman's chanting of the creation story. A similar if forgotten purpose was contained in the tradition of reciting the creation story of John ("In the beginning was the Word") (see also **Christian Creation**) at the end of the Catholic or Anglican mass. Although no longer practiced, this reading indicated a "curing ceremony" in which, with the culture hero, the initiate underwent a process of death and resurrection toward a new beginning.

Creationism

As a theological concept, creationism is opposed to traducianism (the theory that one's soul is inherited from one's parents). Creationism asserts that an original soul is created by God for each human being.

In particular opposition to the Darwinian theory of evolution (see also **Evolution**), creationism is the theory that ascribes the creation of the world—including plants, animals, and especially humankind—to God, as described in the Book of Genesis in the Bible.

Creek Creation

See Yuchi Creation.

Crow Creation

The Absarkoes (Sparrow Hawk), a Plains Indians group wrongly called the Crows, have many versions of a creation myth that features the familiar Old Man **Coyote,** the trickster-creator (see also **Trickster**). It is essentially an **earth-diver creation** myth with elements of creation by the creator's breath, creation from clay, and creation from the creator's words (see also **Creation by Word; Creation from Clay**). In this version of the myth, the role of creator and trickster is played not only by Old Man Coyote, but also by Little Coyote, who acts as the creator's trickster-assistant.

Once there was only water and Old Man Coyote. "I wish I had someone to talk to," said Coyote, and when he turned around he found two red-eyed ducks. "How about diving down to see if there is anything under the water," he said to them. The first duck dove and stayed under for so long that Coyote thought he was dead. After a while, though, he came back and said he had hit bottom. On a second dive he found a root; on a third dive he found a lump of earth.

Coyote was pleased and announced that he would make a place to live using the mud. When he breathed on it, it grew and grew until it was the earth. Coyote then planted the root that the duck had brought up, and this started the plants and trees growing.

"Isn't this beautiful?" Old Man asked the ducks. They said it was, but that it needed valleys, hills, mountains, rivers, and lakes. Coyote made these, and the ducks praised him for his creative talent and cleverness. Coyote was pleased by their reaction, and when they complained of a lack of companions, he took clay and formed people and then more ducks.

Coyote was not pleased, however. "I have made only males," he cried. Instantly he made females, and the people and ducks were happy. They had a good time and multiplied.

One day Old Man Coyote came across a little version of himself, Little Coyote. "Where did you come from, Little Brother?" he asked. "I don't know, Big Brother," Little Coyote answered. "I'm just here."

"Well, Little Brother, I am Old Man Coyote, and I made everything you can see."

"But you need more animals," said Little Coyote. "You've only made ducks and people."

"You're right," said Old Man. Then, as he named animals, they came into being—bear, elk, deer, antelope. The bear complained that the new animals were bored, even though Coyote had made males and females: "That male-female thing gets boring, too, after a while," said Bear.

So Coyote made Prairie Chicken out of Bear's claw, and it taught the other animals how to dance. Bear went on complaining, however, asking for one thing or another and finally claiming that only he should be allowed to dance because he was so big and grand. When Old man warned him not to be so arrogant, especially to his maker, Bear claimed he had made himself. This made Old Man angry, and from that moment Bear was made to live away from the other animals in a den and to sleep all winter so he would be out of the way.

Everything went fine until Little Coyote suggested that the people Old Man had made early in the creation were doing poorly. Old Man then showed these people about making tipis, fires, and weapons and also taught them how to hunt. He gave weapons only to the people, because only they were slow and unprotected by fur, feathers, or claws.

Then Little Coyote did something bad. He suggested to Old Man that he give the people different languages so they would misunderstand each other and use their weapons in wars. He convinced the creator that people would thrive as warriors, horse thieves, and woman thieves. He talked about heroism, war dances, and songs of heroic deeds; he talked of honor. Old Man did what Little Coyote said, and the people had different languages and made war on each other. Some became heroes and chiefs, and some became horse thieves and woman thieves.

Source: Erdoes, 88–93.

Cuebo Creation

The Cuebo Indians of Columbia say the world was always here—that no creator could have made the world. As for the Cuebo themselves, they were here before any other people, and they were born of the rocks in the Vaupas River. At first they were anaconda snakes, but when they molted they became the people and settled where they live now.

The Cuebo do have gods—a god of the dead, who lives in the sky, and the god Quwai, who created some things, primarily fish for the people to eat. Quwai also taught the people how to weave and how to mourn the dead. There is also the god Avya, who is both the sun and the moon.

Source: Leach, 125–126.

Cupeno Creation

This *ex nihilo* Coyote myth is a fragment from the Cupeno tribe of California (see also **Creation from Nothing**).

In ancient times there was darkness and emptiness everywhere, but in space there hung a bag that opened and released Coyote and Wild Cat. The two could not agree on which had come first. The argument was decided when it was Coyote's voice that was heard first by the unformed humans who lived in the void. When they heard Coyote, they rose up and sang the sacred songs, and since then Coyote has been important to them—much more so than Wild Cat.

Source: Sproul, 242.

Dahomey Creation

See Fon Creation.

Death as Creator

The depiction of death as a creator is unusual but not altogether illogical. Creation is, by definition, death-defined; that is, life as we know it depends on deterioration (the gradual progress of disorder) and death as parts of the natural cycle that includes procreation and growth. Creation would be static if there were no death and no procreation. It would be suffocating and self-defeating if there were procreation and no death. Death is, in a sense, the author of all things, literally, the "fertilizer" of creation (see also **Kono Creation**).

Descent into Water

In many creation myths the primal waters are the form taken by chaos, the basic, unformed material of creation (see also **Primal Waters**). The waters have an analogy in the maternal waters in which the foetus is formed and nourished. In **earth-diver creation** myths, an animal enters the maternal waters as a midwife of sorts and brings back the mud or clay of creation. The descent into water, then, is a search for life in the potentially creative womb of the Great Mother herself.
Source: Long, 188–190.

Dhammai Creation

The Dhammai are non-Hindu people of northeastern India (see also **Indian Creation**). Their world parent creation proclaims the essential sacredness of the earth (see also **World Parent Creation**).

Before there was anything, Shuzanghu lived alone with his wife, Zumaing-Nui. They lived on high and were tired of having no place to set their feet. Zumaing-Nui got her husband to promise to be faithful and loving to her in return for her solving their problem. Shuzanghu agreed and made love to his wife.

In time the couple gave birth to a girl (Earth) and a boy (Sky). Since there was nothing to support their weight, however, they fell down and were swallowed by Worm below.

Upset by what had happened, the first parents set a trap for Worm before their next child was born. When they caught him, they split his body open and found their children, Earth and Sky. So the lower part of Worm's body became our earth and the upper part our sky. As for Sky and Earth, they became husband and wife and gave birth to gods, who were mountains. Then they separated, but not before producing two frogs who married and made the first humans. These humans were covered with thick hair, but they married and made the people as we know them.
Source: Sproul, 195–196.

Dieguenos Creation

The creation myth of this southern California tribe contains the world parent motif of the separation of earth and sky (see also **Creation from Division of Primordial Unity; World Parent Creation**) like that of the Egyptian story of Geb and Nut or the Greek myth of Uranos and Gaia. It states that when the creator, Tu-chai-pai, made the world, the earth was female and the sky was male. The sky came down over the earth, which was only a lake covered with rushes, and the creator and his brother were cramped between them.

Tu-chai-pai took tobacco in his hands and rubbed it and blew on it three times. With this, the sky rose a bit over the earth. When both brothers blew the tobacco and said the magic words, the sky went all the way up.

Now the creators made the four directions. Later, since they knew that people were coming, they made hills, valleys, and lakes. They made forests for wood so the men could build things.

Tu-chai-pai announced that it was time to make people. He took mud and made the Indians first and then the Mexicans. It was more difficult making women than men. The creator told the people that they would not have to die but that they would have to walk all the time. He did give them the sleep that could make them still at night. Otherwise, he said, they should walk towards the light in the east.

The people found the light of the sun and were happy. The creator's brother then made the moon. The people were told that whenever the moon got small and seemed to die, they must run races. After they did this, the creator was finished, but he continued thinking for a long time.
Source: Erdoes, 156–157.

Divine Woman as Creator

In some creation myths (see also **Gnostic Creation; Huron Creation**), the creator is female (see also **Great Mother as Creator**). The logic of this is contained in the obvious role of the female as the birth-giver. In pre-patriarchal times, when it seems likely that goddesses rather than gods reigned as the supreme beings, there would have been numerous myths of female creators. With the emergence of patriarchal cultures, these female creators would have been replaced by male figures. There are remnants of the older point of view in the creation myths of some patriarchal societies, for instance, Gaia in the Greek creation (see also **Descent into Water; Primal Waters**).

Djanggawul Creation

Djanggawul is the mythical founder of an aboriginal cult named after him. As in the case of many origin myths, the places and practices referred to are well known to the local people—in this case those of north-eastern Arnhem Land in Australia. The instance of brother-sister incest in the creation of human beings is a common motif among the aborigines (see also **Incest in Creation Myths**). The theme is also present in the creation myths of many other cultures, including, for instance, Greek and Native American traditions.

In the beginning everything in nature existed except for humans. The prehuman ancestors of humans did exist, and these were called the Djang-gawul. They created *ex nihilo* by way of the aborigine Dreaming. They possessed sacred thoughts (see also **The Dreaming**) as well as the magical

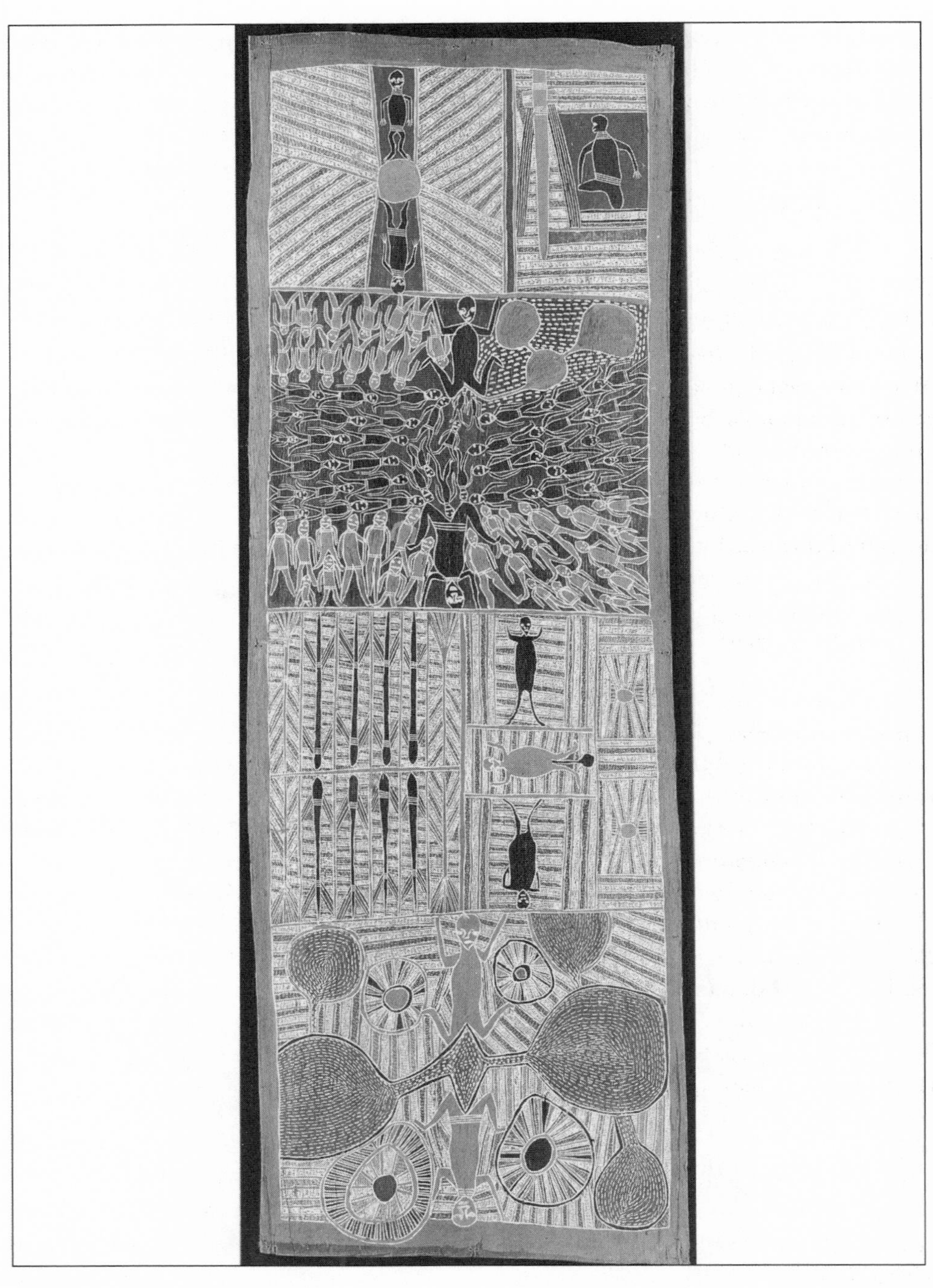

An artist of the Aboriginal Yolngu/Dhangu-language group included himself, upper right, in his representation of the Djanggawul creation story. Told in four panels, or registers, the story begins with two sisters, Miralaidja and Bildjiwuraroju, at a spring, upper left. The second and fourth panels show them giving birth, while the third panel represents trees planted by the Djanggawul, the sisters with their brother, who was their husband, and a rising and setting sun.

objects they carried in a bark canoe to the various parts of the country they wished to populate. There were three of these beings—Djanggawul himself and his two sisters, Bildjiwraroiju and Miralaidj. Djanggawul had a very long uncircumcised penis decorated with notches. The sisters had long clitorises. The sex organs of all three dragged along the ground leaving sacred markings.

Wherever the ancestors beached their canoe they left children made by the brother and the older sister, and later the younger sister as well. They conceived the children in the normal way, but it was necessary for Djanggawul to lift the long clitorises of his mates to do so. Wherever they stopped they also left "dreamings" in the form of objects, sacred stories, and ceremonial traditions. Their sacred sex organs were central to the ceremonies and were represented by decorated poles.

Eventually the Djanggawul came to Jelangbara, the holy destination of their journey and the center to this day of their cult in northeastern Arnhem Land. They made camp in what is now a sacred waterhole made by the Djanggawul when they inserted a sacred pole into the ground (see also **World Tree**). Today a spring flows from the hole. Some say that when they got to Arnhem Bay they instituted the practice of circumcision. In any case, wherever they went they established their cult and left children who would later act together as husbands and wives, and so it was that the human beings of that area came to be.
Source: Long, 234–243.

Dogon Creation

The cosmogony of the Dogon people of modern Mali provides a good example of **incest in creation myths.** It has been suggested by many anthropologists that such incest stories are an attempt to support particular kinship systems. Such is the case with the Dogon cosmogony. The Dogon are a patrilineal people (ownership passes through the male line). If possible, a man will marry his first cousin by his maternal uncle after having intercourse with his future wife's mother. This is because normal incest taboos prevent intercourse with the mother—the ideal spouse—and substitute the aunt/mother-in-law. The Dogon creation—a primal egg myth—validates this reenactment of mythical incest (see also **Creation from a Cosmic Egg**).

The people, according to this myth, are ultimately descended from two sets of twins that were individually androgynous. Since every human being

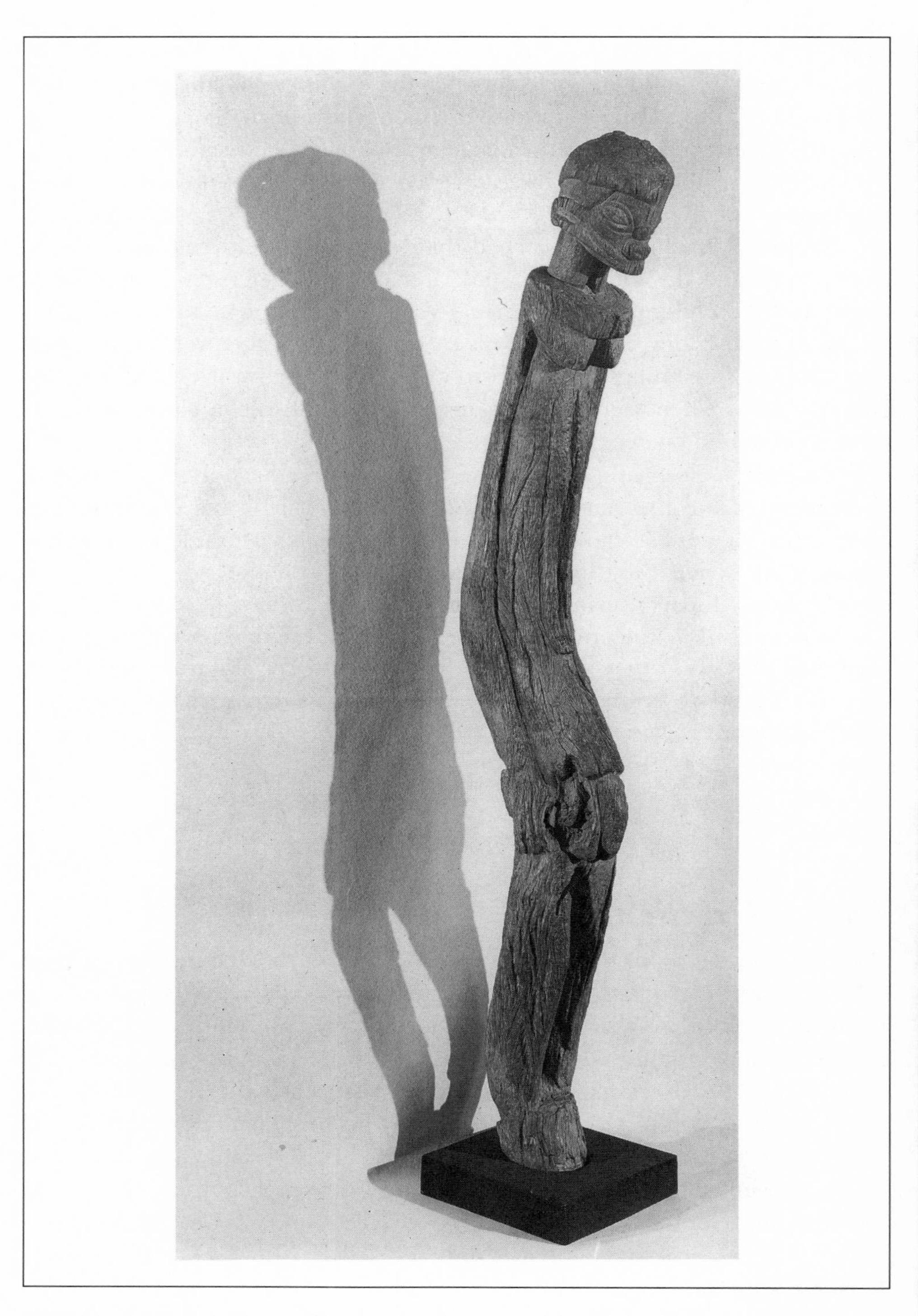

The Dogon of Africa's Sudan region attribute their creation to two sets of twins called Nummo. This carving, collected in the modern West African nation of Mali, represents one of the male twins.

is descended from these twins, every human being is, in a sense, a descendant of a mother and a twin brother or a father and a twin sister. In a complex religious sense, brothers and sisters are seen as parents of each other's offspring.

In the beginning, the world egg was shaken by seven huge stirrings of the universe. It divided into two birth sacs, each containing a set of twins who were fathered by the supreme being, Amma, on the maternal egg. In each placenta was a male and a female twin, but each twin contained both the male and female essence.

By some fluke, a male twin called Yorugu broke out of one of the placentas before the proper time, and the piece of the sac from which he broke forth became the earth. When he tried to go back to the egg to retrieve his twin, she had disappeared. In fact, she had been placed in the other placenta with the other set of twins.

Yurugu went back down to the new earth and copulated with it—his own maternal placenta—but did not succeed in creating people. Seeing this, Amma sent the other twins down to procreate, and so it is that humans came from the original joining of brother, sister, and cousin twins.

The Dogon people of Mali and Upper Volta see creation as a process marked by sacred revelations. The first revelation is nature itself, which speaks through the sounds of the grasses covering the earth. The second is order, symbolized by weaving. Order resulted in humans choosing to live in communities. The third revelation is the granary and the drum. The granary is to the community what the earth is to the cosmos and the stomach is to the individual. The drum is the people's means of communication.

Earth itself was created by the god Amma, who then raped her. Through this act of violent incest came the Nummo twins, who brought order to creation by making sense of opposites.

This is the beginning of a partly *Deus Faber* version of the Dogon creation as told to ethnologist Marcel Griaule by the Dogon elder, Ogotemmeli (see also **Creation by *Deus Faber***). It also contains an "explanation" of female circumcision (see also **Creation Myths as Explanation**).

> Ogotemmeli, seating himself on his threshold, scraped his stiff leather snuff-box, and put a pinch of yellow powder on his tongue. "Tobacco," he said, "makes for right thinking."
>
> So saying, he set to work to analyse the world system, for it was essential to begin with the dawn of all things. He rejected as a detail of no interest, the popular account of how the fourteen solar systems were formed from flat circular slabs of earth one on top of the other. He was only pre-

pared to speak of the serviceable solar system; he agreed to consider the stars, though they only played a secondary part.

"It is quite true," he said, "that in course of time women took down the stars to give them to their children. The children put spindles through them and made them spin like fiery tops to show themselves how the world turned. But that was only a game."

The stars came from pellets of earth flung out into space by the God Amma, the one God. He had created the sun and the moon by a more complicated process, which was not the first known to man but is the first attested invention of God: the art of pottery. The sun is, in a sense, a pot raised once for all to white heat and surrounded by a spiral of red copper with eight turns. The moon is the same shape, but its copper is white. It was heated only one quarter at a time. Ogotemmeli said he would explain later the movements of these bodies. For the moment he was concerned only to indicate the main lines of the design, and from that to pass to its actors.

He was anxious, however, to give an idea of the size of the sun.

"Some," he said, "think is is as large as this encampment, which would mean thirty cubits. But it is really bigger. Its surface area is bigger than the whole of Sanga Canton."

And after some hesitation he added:

"It is perhaps even bigger than that."

He refused to linger over the dimensions of the moon, nor did he ever say anything about them. The moon's function was not important, and he would speak of it later. He said however that, while Africans were creatures of light emanating from the fullness of the sun, Europeans were creatures of the moonlight: hence their immature appearance.

He spat out his tobacco as he spoke. Ogotemmeli had nothing against Europeans. He was not even sorry for them. He left them to their destiny in the lands of the north.

The God Amma, it appeared, took a lump of clay, squeezed it in his hand and flung it from him, as he had done with the stars. The clay spread and fell on the north, which is the top, and from there stretched out to the south, which is the bottom, of the world, although the whole movement was horizontal. The earth lies flat, but the north is at the top. It extends east and west with separate members like a foetus in the womb. It is a body, that is to say, a thing with members branching out from a central mass. This body, lying flat, face upwards, in a line from north to south, is feminine. Its sexual organ is an anthill, and its clitoris a termite hill. Amma, being lonely and desirous of intercourse with this creature, approached it. That was the occasion of the first breach of the order of the universe.

Ogotemmeli ceased speaking. His hands crossed above his head, he sought to distinguish the different sounds coming from the courtyards and roofs. He had reached the point of the origin of troubles and of the primordial blunder of God.

"If they overheard me, I should be fined an ox!"

At God's approach the termite hill rose up, barring the passage and displaying its masculinity. It was as strong as the organ of the stranger, and intercourse could not take place. But God is all-powerful. He cut down the termite hill, and had intercourse with the excised earth. But the original incident was destined to affect the course of things for ever; from this defective union there was born, instead of the intended twins, a single being, the *Thos aureus* or jackal, symbol of the difficulties of God. Ogotemmeli's voice sank lower and lower. It was no longer a question of women's ears listening to what he was saying; other, nonmaterial, eardrums might vibrate to his important discourse. The European and his African assistant, Sargeant Koguem, were leaning towards the old man as if hatching plots of the most alarming nature.

But, when he came to the beneficent acts of God, Ogotemmeli's voice again assumed its normal tone.

God had further intercourse with his earth-wife, and this time without mishaps of any kind, the excision of the offending member having removed the cause of the former disorder. Water, which is the divine seed, was thus able to enter the womb of the earth and the normal reproductive cycle resulted in the birth of twins. Two beings were thus formed. God created them like water. They were green in colour, half human beings and half serpents. From the head to the loins they were human: below that they were serpents. Their red eyes were wide open like human eyes, and their tongues were forked like the tongues of reptiles. Their arms were flexible and without joints. Their bodies were green and sleek all over, shining like the surface of water, and covered with short green hairs, a presage of vegetation and germination.

These spirits, called Nummo, were thus two homogeneous products of God, of divine essence like himself, conceived without untoward incidents and developed normally in the womb of the earth. Their destiny took them to Heaven, where they received the instructions of their father. Not that God had to teach them speech, that indispensable necessity of all beings, as it is of the world-system; the Pair were born perfect and complete; they had eight members, and their number was eight, which is the symbol of speech.

They were also of the essence of God, since they were made of his seed, which is at once the ground, the form, and the substance of the life-force of the world, from which derives the motion and the persistence of created being. This force is water, and the Pair are present in all water: they *are* water, the water of the seas, of coasts, of torrents, of storms, and of the spoonfuls we drink.

Ogotemmeli used the terms "Water" and "Nummo" indiscriminately.

"Without Nummo," he said, "it was not even possible to create the earth, for the earth was moulded clay and it is from water (that is, from Nummo) that its life is derived."

Reprinted from Marcel Griaule, *Conversations with Ogotemmeli,* Oxford: Oxford University Press, 1975, 16–40.

Sources: Middleton, 70; Sproul, 49–66.

The Dreaming

The Dreaming is a theme used consistently in the creation myths of Australian aborigines. Much more complex than the traditional understanding of dreaming common to all people, it refers to a "Dreamtime" of the distant past when deities performed "walkabouts," creating people and sacred places and establishing clans, their totems (spiritual relationships with particular animals), and their socio-religious systems, including taboos. The Dreaming, then, is the given tribe's spiritual and original history, the process by which it and the world around it was created. Every aboriginal carries something of the original Dreaming within himself or herself (see also **Arandan Creation; Djanggawul Creation; Ngurunderi Creation**).

Dyak Creation

The Dyak are an aboriginal Indonesian people of Borneo. They say the sun and the moon were created by the Supreme Being out of sacred clay that is found deep in the earth. They emulate the creator by making sacred vessels out of this clay, vessels that ward off evil spirits (see also **Creation from Clay; Earth-Diver Creation**).
Source: Olcott, 31.

Earth Mother

The Earth Mother takes many forms in creation myths. She is sometimes the generating earth itself, sometimes the unseeded primary waters out of which creation will emerge when affected by the male Supreme Being. In emergence myths the other world is her womb, from which the people emerge. The Earth Mother is a major figure in the world parent type of creation (see also **World Parent Creation**). Typically, everything begins with a primordial union between the Earth Mother and the Male Force (see also **Greek Creation; Indian Creation; Shoshonean Creation; Maori Creation; Okanagon Creation**).

Earth-Diver Creation

This is one of the common forms of creation. The Supreme Being typically sends an animal—a duck, a turtle—into the primal waters (see also **Descent into Water**) to find mud or clay with which to form the earth. The waters can be seen as the unformed female principle, and the diver is the creator god's emmisary into that principle, out of which will come cosmos. Perhaps the emissary can best be seen as the creator's spirit or soul, an equivalent of his breath in creation myths where he breathes life into the world, or his word or thought when he speaks or thinks it into existence. It also might be said that the earth-diver myth suggests a parthenogenic creation from the maternal waters. Whatever the precise meaning, the earth-diving act suggests that the essence of the beginning, whether in reference to psychology or cosmogony, is in the very depths. It

has been suggested that the earth-diver's descent is related to the mythological descent into the underworld or the psychological return to the womb. These myths are particularly common among Native Americans, but are found elsewhere as well (see also **Central Asiatic Creation; Huron Creation; Indian Creation; Iroquoian Creation; Maidu Creation; Negritos Creation; Pohonichi Miwok Creation; Rumanian Creation; Siberian-Tartar Creation; Yokut Creation**).
Sources: Long, 191; Weigle, 71.

Efe Creation

The Efe people of Zaire tell a story that suggests the influence of missionaries who taught the creation story of Genesis (see also **Hebrew Creation**), but the Genesis myth is clearly assimilated into an older story. We note, for instance, that the moon, a female figure, helps the Supreme Being in his creation of humans. We also note the familiar element of **creation from clay.**

With help from the moon, the Supreme Being made a man, Baatsi, out of clay, which he covered with skin and filled with blood. He made a woman, too, and commanded the man to make children with her. "Only be sure to obey one rule," he said. "Do not eat of the Tahu Tree."

So it was that Baatsi fathered many children and his children fathered many more children, and everyone obeyed the rule. When they got old and tired they simply went happily to heaven.

Everything went on this way until a pregnant woman with a craving for tahu fruit convinced her husband to break some off for her. Naturally the moon saw the man picking the fruit in the dark and she told her co-creator.

Because of what the man and the woman did, we now all suffer the punishment of death.
Source: Beier, 63.

Efik Creation

In Nigeria the Efik tribe has a creation myth that reflects local eating traditions as well as patriarchal control over children, a control undermined by the more realistic approach of women and the natural drives of all people to make their own lives. Like many patriarchal creation myths, it blames women for human troubles.

The creator, Abassi, after he had created two humans, feared their ambition. Had it not been for the intercession of Atai, the god's wife, they would not have been allowed to live on earth. Having given in to his wife, Abassi kept control over the people by insisting that they take all their meals with him. They were forbidden to grow or hunt for food, and they were forbidden to procreate. They had to keep their minds on Abassi. Each day a bell rang to call the human couple up to Abassi's table for meals.

Then the woman began growing food in the earth, and the people liked the food and stopped showing up in heaven for meals. Soon the man and woman worked the fields together, and before long one thing led to another and there were children. The man tried to hide the children, but Abassi saw the children and he blamed Atai for not recognizing his fears about the humans in the first place. "Look," he said, "they are making their own food and they are procreating, and they have forgotten all about me; see what you have done?"

"Don't worry," said Atai. "I will not let them take more power." She sent down death and discord to keep the people in their place.
Source: Beier, 60.

Egyptian Creation

The civilization of ancient Egypt is among the longest lasting and religiously complex in the history of humanity. There is a prehistoric period about which not much is known. Then in about 3000 B.C.E. we have the beginning of the so-called Early Dynastic period, which is commonly divided into the first and second dynasties, the first beginning with the union of Upper and Lower Egypt under the rulership of Memphis, a city in the north, near what is today Cairo. It was during this time that both writing and mythological systems were firmly established. The high god during the first dynasty was Horus in his form as Falcon. In the second dynasty, beginning in about 2850 B.C.E., the union collapsed, and Seth became the high god of the southern region, or Upper Egypt. The Old Kingdom, comprising the third through sixth dynasties, covers the period from about 2780 B.C.E. to about 2250 B.C.E. and is marked by the domination of the northern religious center, near Memphis, called Heliopolis. This is a period of great pharaonic power, the great pyramids, the greatest Egyptian art, the high god Atum or Ra, and the emergence of the cult of the resurrection godking Osiris and his wife Isis. It is also the period of the sacred Pyramid Texts, from which we get most of our information regarding the early Egyptian creation myths. A time of anarchy followed during the seventh through tenth dynasties, but the period is notable for its literary activity and the so-called Coffin Texts, which also supply information on the cosmogonies. During this time the southern city of Thebes (now Luxor) grew in power. That power was solidified and Upper and Lower Egypt reunified in the Middle Kingdom era, covering dynasties eleven through thirteen and the period of the high god Amun or Amun Ra. Amun reigned during an intermediate period of four dynasties leading up to the New Kingdom era, which began in 1580 B.C.E. and was marked at the end of the eighteenth dynasty by the religious rebellion of the monotheist sun pharaoh, Akhnaton. During the next three dynasties, beginning in about 1320 B.C.E. and including the rule of Tutankhamen (King Tut), the power of Amun was restored.

Egyptian creation takes many forms, depending on the period and the religious center in question. In prehistoric times it seems clear that a great goddess, sometimes called Nun, reigned supreme and was responsible for creation out of herself. It was said that she gave forth Atum, who then created the universe. There are remnants of this female creative power in such figures as Hathor, Nut (or Neith), and Isis, but by the time of the Pyramid Texts, in which we find the Old Kingdom creation myths, the male force has achieved dominance.

In spite of a constant development over the centuries, certain aspects of an Egyptian creation myth can be said to be relatively constant. These include a source of all things in the primeval waters, themselves a remnant of the Great Mother, and the presence of an Eye, the sun, that creates cosmos within the chaos of the surrounding waters. The sun, whether Atum, Ra, or Ptah, is also associated with a primeval mound or hill, much like the little fertile mounds left by the receding Nile after the annual floods and perhaps like the early sun coming over the horizon. The mound was symbolized by the great pyramids. The people of Heliopolis said their city was the primal mound, the center of creation. The **primal mound** is also equivalent to the clump of earth that is brought up from the primal waters in so many **earth-diver creation** myths.

At Heliopolis, over the centuries Atum took many forms, rather like the Indian concept of Brahman. Atum or Ptah was the original god; Khepri (spelled in various ways—for example, Khoprer) is Atum made visible, and Ra is god as the sun. The Pyramid Texts tell us that Atum existed alone in the universe and that he created his brother and sister, Shu (air-life) and Tefnut (moisture-order) *ex nihilo* (see also **Creation**

from Nothing) by masturbating or, as some texts claimed, by expectorating (see also **Creation by Secretion**). In some places the original god as Khepri, the morning sun, was said to have created himself by word—by calling out his own name (see also **Creation by Word**).

Shu and Tefnut, in a sacred incestuous act to be repeated for centuries by Egyptian pharaohs (god-kings), produced the god Geb (earth) and the goddess Nut (sky) (see also **Incest in Creation**). All of this was watched over by the non-interfering Eye, the original god. Geb and Nut were the parents of Osiris and Isis, Seth and Nephthys, and the older Horus. Osiris and Isis would later produce the boy Horus. From the children of Geb and Nut came all the children of Egypt.

The best known and most frequently depicted event in the Egyptian creation is the separation of the world parents, Geb and Nut (see also **World Parent Creation**). Nut is typically seen arching as the sky over her prone brother, Geb. As the earth, he longs for the moist gifts of the sky so he may procreate, and frequently he is shown with an erect phallus. The world parents are separated by their father, Shu (air), presumably signifying the necessity of differentiation and order rather than total union or nondifferentiation (chaos) for creation.

An early version of what became the Geb and Nut story says that when they perceived the old age of Ra and suspected his weakness, the people began to rebel against him. Not pleased, Ra held a meeting of his Eye, Shu, Tefnut, Geb, Nut, and Nun (the primeval waters) and told them that he had decided to destroy the people for their arrogance. At Nun's suggestion, Ra sent out his Eye to terrorize the people, and they fled into the desert, most to their death.

Wishing to retire, Ra and Nun asked Shu to place himself beneath Nut and raise her up. When he did so, she became the great sky cow, and the earth formed as Geb. A new creation began.

Among the many forms of the Egyptian cosmogonies is the familiar figure of the cosmic egg (see also **Creation from a Cosmic Egg**), a substitute for the primeval waters or the primeval mound. There were people who believed in the cosmic egg as the soul—perhaps the male soul, Atum or Shu—of the original maternal waters (a kind of ancient Animus, to borrow the Jungian term for the opposite sex projection of the energizing soul of the female).

One variant of the cosmic egg version teaches that the sun god, as primeval power, emerged from the primeval mound, which itself stood in the chaos of the primeval sea.

Sources: Brandon, 14–65; Clark, 35–67; Freund, 79–80; Hamilton, 111–115; Leach, 217–220; Long, 99–101, 183–184; Weigle, 73–75.

Ekoi Creation

The Ekoi live in southern Nigeria. They say that in the beginning there were two gods, Obassi Osaw and Obassi Nsi, who created everything together *ex nihilo* (see also **Creation from Nothing**) until Obassi Osaw decided to live in the sky and Obassi Nsi decided to live on the earth. Obassi Osaw gives light and moisture, but he also brings the pain of draught and storms. Obassi Nsi, however, is a nurturer, and he takes us back to himself when he dies.

One day long ago, the sky god made a man and a woman and placed them on the earth. They knew nothing about food or drink, so Obassi Nsi taught them about planting and Obassi Osaw sent down water from his great blue cloak. The people learned to grow and eat the fruit of the palm tree and to use it for medicine as well. The gods have treated the people well.

Source: Leach, 138–139.

Eliade, Mircea

Mircea Eliade was a scholar of world religions and the original classifier of cosmogonic myths into the basic types. Eliade sees the creation myth—the myth of chaos transformed into cosmos—as the essential human myth, the myth of existence itself.

Elohim

This is one name for the Hebrew creator god; another is Yahweh (see also **Hebrew Creation**).

Enki

Enki was one of the creator gods of the ancient Sumerians in Mesopotamia. Marked by **trickster** aspects, he was, next to the goddess Innana/Ishtar, the most popular of the Sumerian deities. His later Semitic Akkadian name in Assyria and Babylon was Ea (see also **Assyrian Creation; Babylonian Creation; Sumerian Creation**).

The Sumerians of Mesopotamia believed Enki, center, to be one of their creator gods. This modern impression was made from a seal created circa 2100 B.C.

Enuma Elish

Enuma Elish is the name commonly applied to the great Babylonian creation epic. The name is taken from the epic's first line, "When on high (Emuma elish) the heaven had not been named" (see also **Babylonian Creation**).

Eruption Creation Theory

American scientists Forest Ray Moulton and Thomas C. Chamberlin developed the eruption theory of creation early in the twentieth century. Billions of years ago, they said, a large star passed close enough to the sun to raise two huge eruptions. These sun eruptions sent out material that orbited around the sun and formed the nine planets. Our planet is, of course, one of these, and meteors are smaller bits of the original eruptions. The original straying star went off on a huge orbital path (see also **Creation in Science**).
Source: Leach, 18.

Eskimo Creation

At the center of many Eskimo creation myths, especially the **earth-diver creation** type, is the figure of Raven who, like Coyote, is a trickster-creator (see also **Chuckchee Creation; Trickster**). Sometimes Raven throws a spear, like a hunter, into the primeval waters and brings up a clump of dirt that becomes the earth. Sometimes he sends other animals to dive for earth. Raven is not always the central figure of the Eskimo myths, however. There are other important animals, such as Hare and Fox, who argued at the beginning as to whether there should be light or darkness. Sometimes humans are here before anything else.

Some Eskimos tell of two men coming out of the earth after a great flood (see also **The Flood**). The two men lived together as a married couple, and when one became pregnant, the other chanted magical words that resulted in the first man's penis splitting into a vagina, out of which emerged the first child.

There is also a tale of two original girls who left the creator-father and fled to the sky, where to this day they make thunder and lightning with an old skin and a flintstone.

The incest motif (see also **Incest in Creation Myths**) runs through many Eskimo creations. One myth is about a sister who, after putting soot on her hands, discovers by the soot left on her brother's back that it is he who has been visiting her bed each night. Angry, she takes a torch and leaves, followed by her brother, who carries a not-so-bright torch. When they both enter the sky, she becomes the sun and he the moon.

Eskimo myths, like those of the Australian aborigines and many other peoples, tend to be local rather than cosmic. A particularly popular example, which is also a good example of a Raven myth, tells us that in the beginning the first man was asleep in a pod. Eventually he broke out of the pod and fell to the ground fully grown, leaving the pod hanging on the vine. Suddenly Raven came to him and changed himself into a man the way the shamans do when they take off their Raven masks. Raven asked the man where he had come from and was surprised to hear that he had come out of the vine, since he himself had made the vine. He asked the man whether he had eaten anything, and the man said he had drunk something from the mushy ground onto which he had fallen. "That was water," said Raven. "Wait here." Raven became a bird again and flew away. He returned in four days with berries, which he instructed the man to eat and plant. Later Raven made sheep out of mud (see also **Creation from Clay**) and blew life into them by flapping his wings.

Many more men grew in the pods on the original vine, and Raven made lots of animals and plants for them to eat and plant. So the men would not eat everything, he made the animals and plants hard to catch and find, and he made the bear—also out of clay—to frighten him.

Finally, feeling that Man was lonely, Raven went off alone and formed a beautiful figure out of clay and blew life into it with his wings. This was Woman, and Man liked her and soon they had a child. So the world went on (see also **Kodiak Creation; Kukulik Creation; Netsilik Creation; Nugumiut Creation**).

This is a magic creation chant of one group of Eskimos.

Magic Words (after Nalungiaq)

In the very earliest time,
when both people and animals lived on earth,
a person could become an animal if he wanted to
and an animal could become a human being.
Sometimes they were people
and sometimes animals
and there was no difference.

All spoke the same language.
That was the time when words were like magic.
The human mind had mysterious powers.
A word spoken by chance
might have strange consequences.
It would suddenly come alive
and what people wanted to happen could happen—
all you had to do was say it.
Nobody could explain this:
That's the way it was.

Sources: Bierhorst, 58–60; Hamilton, 3–7; Leach, 195–197; Olcott, 25–26.

Etiology

Etiology is the science of causes. An etiological myth is one that assigns causes for a phenomenon. In a sense, then, all creation myths are etiological (see also **Creation myths as Explanation**).

Eurasian Creation

See Finnish Creation.

Eve

Eve is Adam's wife—the first created woman—in the Genesis creation myth of the Bible (see also **Adam; Hebrew Creation**). Archetypically Eve is related to the Greek Pandora and all other original women to whom patriarchal mythologies attribute the beginning of human problems. Eve's name may well be related to the Indian name Jiva or Ieva, the creatress of forms. It also has roots in the many Middle Eastern names for the life force: for example, the Hittite and Aramaean Hawwah and the Persian Hvov.

Evolution

The theory of evolution, developed by Sir Charles Darwin in the nineteenth century and now generally accepted by the scientific community, reveals a continuous process of creation in which all species of animals and plants develop from earlier biological forms by hereditary transference of variations from one generation to another. The natural selection aspect of Darwinian theory postulates development of a given species in relation to the requirements of its environment, noting a phenomenon called survival of the fittest, in which survival belongs to those species that adapt best to the world around them. Evolution is generally discounted by creationists, those who believe in traditional religious creation stories (see also **Creationism**).

F

The Fall of Humankind

A common motif of creation myths is that of the fall from grace of the first humans. Often, as in the case of the **Hebrew creation,** the first humans do something forbidden by the creator. Sometimes they are unsuitable in the form in which they have been created, as in the Quiché Mayan creation myth, or sometimes they have been generally corrupted (see also **Imperfect Creation**). Often the moral fall is followed by a complete fall or cosmic cleansing, usually in the form of a flood (see also **Babylonian Creation; Egyptian Creation; The Flood and Flood Hero; Greek Creation**).

Fang (Fans) Creation

A farming Bantu people called the Fang or Fans live in Africa in what is now Gabon. They tell of Mebere, who made the first human out of clay, but in the form of a lizard. He put the lizard in the sea and left him there for eight days, after which he emerged as a man and said, "Thank you" (see also **Creation from Clay**).

The Fang also have a more detailed *ex nihilo* creation myth (see also **Creation from Nothing**), notable for its emphasis on the fall of the first human (see also **The Fall of Humankind**) and for its vision of God as a trinity.

In the beginning there was nothing but Nzame (God). Nzame is really three: Nzame, Mebere, and Nkwa. It was the Nzame part of Nzame who created the universe and the earth and blew life into it. When he had done that he called Mebere and Nkwa, the other parts of Nzame, to admire his work. "How do you like it?" he asked. "Do I need to make anything else?" Mebere and Nkwa thought for a while and then suggested that Nzame make a chief for the earth. First Nzame named the elephant, the leopard, and the monkey as joint chiefs, but he wanted something better.

So, between them Nzame, Mebere, and Nkwa made a creature in their image. They named him Fam (Power). From Nzame the new creature gained strength, from Mebere he received leadership, and from Nkwa he had beauty. They charged Fam to rule the world and then went back up high where they live.

For a while everything went well; even the first chiefs—the elephant, the leopard, and the monkey—obeyed Fam. Fam grew arrogant, however, too proud of the qualities he had been given by Nzame. He began to mistreat the animals, even the first chiefs. He decided he did not need to worship Nzame; "Let Nzame be where he is; I rule here," he sang.

Nzame was not pleased by this song. "Who is singing?" he called out. "Why not try to find him?" Fam answered rudely, so Nzame brought down thunder and lightning and destroyed everything that was. Only Fam was left, because he had been promised a deathless existence, and Nzame cannot take back his word. This is why, although Fam disappeared, we know he still exists and that he harms us when he can.

Looking down at the parched earth, Nzame, Mebere, and Nkwa decided to fix it up. They applied a new layer of earth to it, and when a tree grew it dropped its seeds and made new trees. Leaves that dropped into water became fish, and leaves that fell on the earth became animals. Soon our earth was reborn as we know it. The old parched earth can be found still if we dig deep enough. We call this coal, and it still burns in fires.

"Now let us make a new Fam," said Nzame; "but this one must know death." The new man was our ancestor Sekume, who made the first woman, Mbongwe, from a tree. The first ancestors were made with both Gnoul (body) and Nsissim (soul). Nsissim gives life to Gnoul and is its shadow. When Gnoul dies, Nsissim does not die; he is the little shiny spot in the middle of the eye. This spot is like the star in the heavens or the fire in the hearth.

Sekume and Mbongwe prospered and produced many children. Their only worry was Fam, who sometimes tunneled up from where Nzambe had left him under the earth and did bad things to them. This is why the people tell their children to be careful what they say lest Fam be listening and bring them trouble.

Sources: Beier, 18–22; Leach, 135.

Farmer's Almanac Creation

According to the *Old Farmer's Almanac,* the world began at 9:00 A.M. on Wednesday, 26 October 4004 B.C.E. This date was determined by an Irish bishop, James Ussher.

Father Creator

The creator is, more often than not, seen as a father. This reflects the reasoning of patriarchal cultures about the proper order of things. The **Sky Father** dominates the world as the earth fathers dominate families and governments. In matriarchal or matrilineal cultures (and in some patriarchal cultures), the creator is apt to be the Great Mother (see also **Mother-Creatrix; World Parent Creation**).

Female Mediator in Creation

The motif of a female assistant to the creator is common, especially among Native Americans (see also **Arikara Creation; Pawnee Creation; Sioux Creation**). Whether she be Corn Mother (Arikaran), White Buffalo Calf Woman (Lakota Sioux), or Star Woman (Pawnee and many others), the mediator works with the people, the way a mother traditionally does with her children, to teach them husbandry and proper ways.

Fiji Islands Creation

The Fiji Islanders in the Melanesian Islands of the western Pacific believe that in the beginning there was only water and twilight everywhere and the island place of the gods. No one knows which island this is, but it floats around the edge of the world and can sometimes be seen at sunrise.

The Fiji creator is the serpent god Ndengei. He was the head god of all those original Fijian gods of the **creation from chaos,** the ones who were here before the Polynesians and Europeans brought their gods. The people say that when Ndengei sleeps it is night, when he rolls over there is an earthquake, and when he wakes up it is day. Ndengei's son, Rokomautu, did the actual earth-diver creating of the Fiji Islands (see also **Earth-Diver**

Creation). He scooped them up from the bottom of the sea. As for Ndengei, he pretty much stays away, but the people pray to him anyway.
Source: Leach, 176.

Finnish Creation

A **creation from chaos** story of the Finno-Urgic peoples is contained in the Finnish national epic, the *Kalevala*, especially in the first two runes or sacred chants. Central to the myth, which also has earth-diver and cosmic egg aspects, are the familiar watery chaos and the first mother (see also **Creation from a Cosmic Egg; Earth-Diver Creation**).

In the beginning there were the primeval waters and Sky. Sky's daughter was Ilmatar, who lived alone and one day drifted down to the waters to rest. There she floated and swam for 700 years longing for more life. There was a day when, floating on her back with one knee raised out of the water, she noticed a beautiful bird, a teal, fluttering over the seas in search of a resting place to make a nest. Ilmatar, the Mother of the Waters, then raised her knee farther so the teal thought it was a dry island. The bird made a nest there and laid six golden eggs and one iron one.

The little teal sat on her nest warming the seven eggs and also heating the knee of Ilmatar. The heat became so great that it began to burn the uplifted knee. Finally, Ilmater could not stand it and she jerked her knee into the water to cool it. This dislodged the eggs, which fell into the water and were shattered by the wind and waves.

Then something full of wonder came to be. From the lower part of one of the eggshells land developed, and from the top was made the sky as we know it. The moon and the stars came from the egg whites, and the yolk became the sun.

After several hundred more years of floating and admiring the results of the broken eggs, Ilmatar began acting on her urge to create. Full of the power of life, she only needed to point to create cliffs and inlets. Her footprints became pools for fish. The movement of her arms made beaches. She made everything that was (see also **Animism**).

One day Vainamoinen was born of Ilmatar. The sea was his father. He swam for years on the sea before landing on a barren island. When he, the first man, stood, he looked into the sky and asked help of the Great Bear in the stars. Help came in the form of a boy carrying seeds, which he scattered at Vainamoinen's command. The seeds became trees and plants and covered the barren land. One seed became the oak, which after many years became so large it brought darkness to the land.

Vainamoinen prayed to his mother for help, and she sent a little man with a copper ax. The man grew into a giant; with his now huge ax he chopped down the oak, and light was restored to the world.
Sources: *Kalevala*; Leach, 239–243.

First Ancestors

These are the first man and woman found in so many myths. They can be created out of each other, as Eve was said to have been created from Adam's rib in the **Hebrew creation** story, or out of clay as in so many other creation myths. Sometimes they are created *ex nihilo*. Typically, in patriarchal cultures, the man is created first and then the woman. See illustration on page 96.

The Flood and Flood Hero

In many creation myths the creator becomes disgusted with the behavior of human beings and cleanses his creation with a great deluge. Often, however, he chooses one human to preserve life for re-creation after the Flood. The flood hero usually takes his wife and several sets of animals with him and rides out the flood in a boat. He represents the positive seed of the original creation, which we hope lies within us all. Whether he is called Ziusudra (Sumerian), Utnapishtim (Babylonian), Noah (Hebrew), Manu (Indian), or Deucalion (Greek), he is the representative of the craving for life that makes it possible for us to face the worst adversities (see also **The Fall of Humankind**).

The Flood occurs, then, as a secondary stage of many creation myths. It is one of mankind's earliest "memories." We cannot "remember" the events of the creation itself, because we were not there, but we were there for the Flood. The persistance of this archetype, expressed so universally in deluge myths, suggests an important aspect of humanity's vision of both its own imperfections and of the possibility of redemption in a new beginning. A microcosmic version of the Flood is to be found in purification ceremonies such as the Christian baptism, in which the initiate "dies" to the old way in the waters of the font and is reborn in Christ. The Flood is a given culture's rebirth from the chaotic maternal waters.

Adam and Eve face one another in the Garden of Eden on a page from the Manafi al-Hayawan *(Book on the uses of animals), compiled in the Persian city of Maragheh (now in Iran) in the late thirteenth century. The city had fallen to Mongols, whose influence lends an oriental cast to the picture.*

Fon Creation

The Fon people are descendants of the first inhabitants of Dahomey in West Africa. Their *ex nihilo* creation myth is told in various ways, the creator sometimes being a Great Mother figure and sometimes a being who is both male and female (see also **Creation from Nothing**).

Most people say the creator was Mawu, the moon, the mother of all gods and people. It is said as well that Mawu can be two beings, Mawu (moon in the sky) or Mawu-Lisa (moon-sun), who is both male and female. Some say that this father-mother creator is different from both Mawu and Lisa, that the creator is Nana-Buluku, who gave birth to Mawu and Lisa and then gave power over creation to them. Nobody really knows for sure, but we are told that Mawu lives in the West and Lisa in the East and that when there is an eclipse they are making love.

There are many popular tales of Mawu as **Mother-Creatrix**. It is said that she created everything, moving from place to place on the back of the rainbow serpent Aido Hwedo. When everything was done, Mawu asked the serpent to coil himself around the too heavy and unsteady earth to keep it stable. Aido Hwedo is still there today, surrounding the earth and holding his own tail in his mouth. Mawu surrounded him with the seas to keep him cool as he supports the heavy world. When he tries to shift his weight once in a while to get comfortable, there is, of course, an earthquake or a tidal wave. One day, they say, he will swallow his tail and the world will topple and come to an end in the sea.

Sources: Leach, 136–137; Sproul, 75–76.

Franz, Marie Louise von

Dr. Marie Louise von Franz is one of the most celebrated followers of Swiss psychiatrist C. G. Jung. She is author of *Patterns of Creativity Mirrored in Creation Myths.* She has concerned herself with the *Deus faber* theme in creation myths, in which the creation of the world has analogies in various arts and crafts: architecture, carpentry, and so forth (see also **Apache Creation; Creation by *Deus Faber***).

The Frost Giant

The giant Ymir (or Imir) is the source of creation in the Icelandic myth. He is an unusual source because he is seen as evil. This is in keeping with the strangely pessimistic view of life contained in Norse myths (see also **Icelandic Creation**).

Fulani Creation

These West African people of Mali say that the world was created from a drop of milk, perhaps suggesting an earlier mother goddess *ex nihilo* creation (see also **Creation from Nothing**). As the story goes now, it is a domino theory creation with what is probably a Christian-influenced ending.

First there was a huge drop of milk. The god Doondari came out of it and made stone. The stone created iron, the iron made fire, the fire created water, and the water made air.

Next, Doondari came down to earth a second time and made a man. The man, however, was arrogant, so the god created blindness to humble the man. Blindness became proud, so Doondari created sleep to defeat blindness, and when sleep got out of hand Doondari created worry to disturb sleep. When worry became too strong he made death to defeat worry. When death became arrogant, Doondari came down for a third time as Gueno, the holy one, and Gueno overcame death.

Source: Beier, 1–2.

Gaia

Gaia (Gaea or Ge) is Mother Earth. She is the oldest of the Greek deities, and in some of the **Greek creation** myths she is the primary creatrix. She was born of chaos itself, and it was she who brought forth Uranus (the heavens) to be her mate.

The following is the Homeric Hymn to Earth.

The mother of us all,
the oldest of all,
hard,
splendid as rock

Whatever there is that is of the land
it is she
who nourishes it,
it is the Earth
that I sing
Whoever you are,
howsoever you come
across her sacred ground
you of the sea,
you that fly,
it is she
who nourishes you
she,
out of her treasures

Beautiful children
beautiful harvests
are achieved from you
The giving of life itself
the taking of it back
to or from
any man
are yours
The happy man is simply
the man you favor
the man who has your favor
and that man
has everything

His soil thickens,
it becomes heavy with life,
his cattle grow fat in their fields,
his house fills up with things

These are the men who govern a city with good laws
and the women of their city,
the women are beautiful
fortune,
wealth,
it all follows
Their sons glory
in the ecstasy of youth
Their daughters play,
they dance in the flowers,
they skip
in and out
on the grass
over soft flowers
It is you
the goddess
it is you who honored them
Now,
mother of gods,
bride of the sky
in stars
farewell:

but if you liked what I sang here
give me this life too
then,
in my other poems
I will remember you

Reprinted from Charles Boer, trans., *The Homeric Hymns,* 2d edition, Dallas: Spring Publications, 1970.

Gaia Principle

The Gaia principle is a scientific hypothesis developed by British scientist James Lovelock and American scientist Lynn Margulis. It takes its name from the Greek earth goddess **Gaia** and suggests that the earth can best be seen as a powerful, self-regulating, living and, in a biological sense, conscious organism. For Lovelock, Gaia (our earth) is a biological "control system" in which humans will play a part only as long as they are useful. There will come a time when Gaia will "eat" her children in what the physicists call "heat death." As for Gaia herself, she will continue to exist until the universe collapses (see also **Creation in Science**).
Source: Lovelock.

Geb and Nut

Geb (male—earth) and Nut (female—sky) are the Egyptian version of the primordial unity that is separated by Shu (the spirit of life) so creation can occur between them. They are best known in the many ancient Egyptian depictions of the naked Nut—often star-spangled—being held up by Shu over the prostrate Geb (see page 102). Geb's penis is sometimes erect, reaching up in creative longing for the object of its desire. It is of interest to note that earth here is associated with the male rather than the female, perhaps because the ancient Egyptian word for sky was feminine and the word for earth masculine (see also **Creation from Division of Primordial Unity; Egyptian Creation; Separation of Heaven and Earth, World Parent Creation**).

The star-spangled ancient Egyptian sky goddess Nut arches over Shu, the spirit of life, who raises his arms to separate Nut from her brother the earth god, Geb, who reclines at Shu's feet.

Genesis

The book in the **Bible** that contains the **Hebrew creation** story.

Germanic Creation

See Icelandic Creation; Norse Creation.

Gilbert Islands Creation

The Gilbert Islanders, Micronesian people of the South Pacific, speak of the eternal one, Nareau, who commanded the sand and water to produce offspring. The myth is a good example of the separation of world parents variety (see also **Creation from Division of Primordial Unity; World Parent Creation**) and includes the element of the god sacrificed to become the world (see also **Babylonian Creation; Creation by Sacrifice**).

One child of water and sand was Nareau the Younger, who called on the others to rise up and live. Because the sky was so heavy on the earth, however, they could not get up. Nareau the Younger therefore killed his father, made the sun and moon out of his eyes, and placed his spine on end on Samoa (see also **Animism**). This became the world tree (see also **Yggdrasil**), and humans were born of it.

The people on Naura Island have an earth-heaven, world parent separation myth in which we are told of the beginning as a time when there was only the sea, in which the god Areop-Enap lived in a mussel shell. It was dark in there, but the god found a big snail and a little snail living with him, so he made the big one into the sun and the little one into the moon. He got a worm to separate the parts of the shells to become the sky and the earth. The worm's sweat made the sea as we know it.
Source: Leach, 183–184.

Gnostic Creation

Gnosticism was a form of mysticism that gained great popularity in the Mediterranean world in the centuries immediately following the death of Jesus. It was a movement influenced by dualism—the struggle between good and evil—as found in Zoroastrianism, Christianity, Buddhism, and perhaps Hinduism as well. A Gnostic was one who possessed gnosis, the special kind of knowledge that would save the soul from the corruption and illusion of the material world.

The most famous Gnostic was Hermes Trismegistus, who lived in Egypt in the second century C.E. In his writings we see a clear parallel with much of the **Christian creation** myth of John with its emphasis on the Word (Logos). The presence of the visionary aspect of Gnosticism is evident, as well, in the Book of Revelation, the last book of the Christian Bible.

In the **creation from chaos** Gnostic story, Hermes tells how Poimandres, the Shepherd, the Nous (Mind) of the Absolute Power, taught him the nature of reality and God. Poimandres begins by giving Hermes a vision of the sacred flame, the light that is Logos, the Word of Nous (Mind of God). He then reveals that he is the light and that the Word of that light is the Son of God.

God, he says, is androgynous, and by taking into itself the Word, becomes the cosmos. God also brought forth the Demiurge, the god of matter who participated in setting the cosmos in motion.

Then Nous created man in his own image and loved him, but man fell in love with the reflection of Nous in the waters, and he fell into the realm of nature—the lower order of the material world. Thus man became both mortal (of the earth) and immortal (of God), and his life is a constant struggle between the dualities, good and evil, the spiritual and the material (see also **Buddhist Creation; Sophia**).

Another Gnostic creation is that of Mani, one of the greatest thinkers of the movement. Mani tells how everything in creation is a struggle between good and evil, light and darkness. First, man was defeated by the powers of Darkness and only partly redeemed by God's messenger, who succeeded in activating creation. Then Darkness created Adam and Eve and taught them to procreate. God followed by sending Jesus to teach man gnosis, the knowledge that can save us. Adam longs for the lost light revealed by Jesus.
Source: Sproul, 142–151.

Great Mother as Creator

There are many examples of the Great Mother as creator (see also **Divine Woman as Creator; Mother-Creatrix**). Such myths occur typically in neolithic goddess cultures—such as those of the Fons, Kagabas, and Sumerians—that stress the fertility of earth as the center of life and civilization (see also **Earth Mother; Gaia**).

Ancient civilizations of the Mediterranean island of Crete attributed the creation of their world to a female goddess. Here, a crowned goddess figurine from Crete raises her arms in benediction.

Greek Creation

Among the oldest creation myths on the Greek peninsula is that of the Pelasgians, dating from the fourth millenium B.C.E. (see also **Pelasgian Creation**). The creation referred to by Homer in the *Iliad* (Book 14) in about the eight century B.C.E. is really a version of that myth. It says that Oceanus and the Titaness Tethys begat the first gods and formed the original world, and that Tethys ruled the sea and Oceanus surrounded the universe.

It is not until the work of Hesiod, who lived in Boetia in the late eighth century B.C.E., that we find a fully developed Greek creation myth. The story is contained in Hesiod's *Theogony* and his *Works and Days.* Note the presence in this story of the theme of the primordial parents, Earth and Heaven (Sky) and their separation (see also **Creation from Division of Primordial Unity; Geb and Nut; Separation of Heaven and Earth; World Parent Creation**). The myth rings particularly familiar to students of Freud's writings on the primal relationships between son and mother and son and father.

It has been suggested by many scholars that the changing of the guard in heaven marks historical changes in the development of Greek culture. According to these theories, in ancient times Greece would have been dominated by an earth-based agricultural society that emphasized the great mother goddesses. In time, however, more warlike patriarchal people invaded the peninsula, bringing with them their thunder-bearing Aryan head god, who became Zeus and put an end once and for all to the ancient struggles between Earth and Heaven, female and male power.

First there was the void, chaos, out of which sprang Gaia, "widebosomed Earth," a firm foundation for Mount Olympus and the gods who would live there. Then came dark Tartarus (Hades), a place deep in the ground; Eros, which is the love that overpowers all; Erebos, the darkness of Tartarus; and Night itself. Air and Day were born of Night by Erebos.

Gaia (Earth) gave birth by herself to Uranos, starry Heaven. He was equal to his mother and covered her completely. Of herself Gaia also brought forth the hills, mountains, valleys, and Pontus, the deep barren sea. After that she lay with her husband-son Uranus and mothered the first gods, the Titans. Some of the best known of these were Oceanus, beautiful Tethys, the one-eyed Cyclopes, three horrid hundred-armed monsters, Hyperion, the earth goddess Rhea, and the terrible Kronos.

From the first, Uranos and his son Kronos hated each other. In fact, most of the children of Uranos and Gaia hated their father. As for Uranos, he hated his children so much that he hid them away in dark places of

Earth (Gaia), causing his mate much grief and pain. In anger at Uranos, she asked her sons for help, accusing Uranos of shameful acts. Wicked Kronos was quick to agree to destroy his father, and he took the great sickle fashioned by Gaia and waited for his unwary parent.

Bringing dark Night with him, Uranos came in lustful passion to beautiful broad-breasted Gaia. When he had laid himself down on her, his son Kronos reached between his parents with the cold sickle and "harvested" his father's parts, flinging them immediately into the wide sea. The drops of blood that fell onto Gaia flowered into the vengeful Furies, the terrible giants, and the graceful Meliaean nymphs. As for the immortal genitals floating on the sea, their seed formed a thick foam around them and out of that foam came the most beautiful of women, who touched first on Cythera and then on the island of Cyprus. This was Aphrodite, goddess of desire, whose follower (and later people said her son) was Eros.

Now Night gave birth to Doom, Death, Sleep, and Dreams. It then bore Blame and Distress, the dreaded Fates, Nemesis, Deceit, Friendship, Old Age, and Strife, who herself mothered Work, Hunger, War, and other miseries.

Kronos now was king over heaven and earth. He raped his sister Rhea, who gave birth to a family of boundless power and beauty (see also **Incest in Creation Myths**). These were the first six of the family of gods that would later be called the Olympians because they lived on Mount Olympus. Hestia was to be goddess of the hearth, Hades god of the underworld, and Poseidon god of the sea. Demeter would take the place of her grandmother, Gaia, and her mother, Rhea, as goddess of earth in a time when humans and gods thought the activities of the heavens more sacred than those of earth. The fifth child of Kronos and Rhea was beautiful Hera, and great Zeus was the sixth.

Evil Kronos had heard from his mother and his emasculated father that he would be overthrown by one of his own, so as each child was born, he ate it. Rhea was horrified at her husband's deed and managed, with the help of her parents, to give birth to her last child, Zeus, in a hidden place on Crete. On Kronos's plate she substituted a stone for Zeus. The deception was successful. The spared Zeus returned in time and led a war against his father and the Titans, eventually establishing the Olympian hegemony.

Zeus ruled supreme on Olympus and married his sister Hera. As king and queen, they produced the blacksmith god, Hephaistos, and the god of war, Ares. Zeus was not a faithful husband. He had children by other goddesses and by mortal women. Among his most powerful offspring were the great goddess Athena, who some say sprang fully armed from Zeus's head; the messenger god, Hermes; the god of prophecy and light, Apollo; his sister, the huntress Artemis; and the mysterious god who dies and returns to life, the holy Dionysos. In the heyday of the Olympians it was said that Zeus, not the severed genitals of Uranos, had fathered the beautiful but dangerous Aphrodite and that she, not chaos, gave birth to Eros.

As for the creation of man, there are many Greek tales. Hesiod says that a golden race of humans was created in the time of Kronos's reign. This race was later hidden away in the earth to become benificent spirits. Then a Silver Age race of somewhat foolish and irreverent humans was created by the Olympians. They angered Zeus and were hidden away in the earth as underworld spirits. Then Zeus made Bronze Age people who were powerful, warlike, and self-destructive. These men were also buried in the earth, in Hades itself. To replace them, Zeus made a race of heroes, whom we know from the stories of the blind Homer. These heroes passed like the other races before them, but they live forever in a far-off place called the Blessed Isles.

The race that lives here now, says Hesiod, is that of Iron. What a terrible fate it is to be of this race—to toil, suffer, and die.

There are people who say the great Titan Prometheus, who sided with Zeus in the War in Heaven and who was later betrayed by him, created mankind out of water and clay, and that he also gave humans fire.

A more philosophical Greek creation was that of the sixth century B.C.E. religious system called Orphism, after the mythological poet-musician Orpheus, who lost his love, Eurydice, in the underworld. The Orphics, like the Zoroastrians of Persia (Iran), emphasize questions of duality and the life of the soul. Their creation myth is of the cosmic egg variety (see also **Creation from a Cosmic Egg**).

The silver cosmic egg was created by Time and burst to give birth to Phanes-Dionysos, the androgynous first god, who was the source of all things, mortal and immortal. Phanes created a daughter, Nyx (Night), out of himself. With Nyx he produced the first gods told of by Hesiod: Gaia, Uranos, and the rest (see also **Orphic Creation**).

Sources: Leach, 234–235; Morford, 29–67; Sproul, 157–169.

Guarani Creation

The Guarayu-Guarani Indians of Bolivia say that in the beginning, after water and bullrushes came about, there was Mbir the worm slithering about in the rushes. Eventually he became a man, and in that shape he created the world as we know it out of chaos (see also **Creation from**

Chaos). Mbir was also called Miracucha, and there were two other gods. One was the sun, Zaguagua, who was so brightly decorated that nobody could look at him until he was far down in the sky in the evening. The other god was Abaangui, who after trying hard to become human finally succeeded, but with a nose so huge that he cut it off. It flew into the sky and became the moon.

The first Guarayu-Guarani ancestor was Tamoi, or Grandfather. He showed the people how to do all the things they do—planting, gathering, hunting, fishing, and making beer, bows and arrows, and fire. When he had done all this Tamoi went away into the west, leaving his wife and child behind as sacred rocks.

It could be that Grandfather was really Mbir, because when people die, their souls go west, where Tamoi went. On the way there they always meet the Grandfather of Worms, who gets huge and blocks the path if the soul has come from a bad person.

The people still pray to Tamoi for good things and do songs and dances for him, so he is probably really Mbir after all.

Relatives of the Guarayu-Guaranis, the Apapocuva-Guarani Indians of Brazil, tell a somewhat different story. They say that in the very beginning, in the dark nothingness, bats fought each other without stopping and that Our Father, who was the sun, entered the darkness and created the earth and propped it up on the Eternal Cross. He then made a woman called Our Mother, who produced twins—Our Older Brother and Our Younger Brother—who continued the work of creation.

Our Father also had a son called Tupa (Thunder). Our Father stays far away from the world and leaves the details of creation to his children. Most people think the elder brother is really the sun, who brings heat and growth, while the younger brother is the pretty, useless moon, who in the old days went around making mistakes that Our Elder Brother had to repair.

It was Our Elder Brother who gave men fire and taught them the dances. The people fear the return of the ancient bats and think the world is coming to an end anyway.

Source: Leach, 119–122.

Guinean Creation

See Kono Creation.

Haida Creation

The Haida Indians of British Columbia and Alaska share the north country fascination with the trickster-creator **Raven,** or Old Man (see also **Trickster**). The creation itself is essentially *ex nihilo* (see also **Creation from Nothing**).

There was a time when only the sea existed and Raven, who was a god then, was flying over it. He spied a tiny island below him and he commanded it to become earth, and it did. When the new earth had grown a lot, Raven cut it up, making Queen Charlotte Island out of a small piece and the rest of the world out of a large one.

On one of his walks around the world he heard a sound coming from a small clamshell and he saw a small face there. After a lot of coaxing, five little faces appeared and then five little bodies. These clamshell-beings were the first people.

The Haidas also have funny stories about Raven stealing the sun and about him getting himself into all sorts of trouble. Even though he was the creator, he was a trickster and, therefore, sometimes very foolish—like us.

Source: Leach, 51–53.

Hard Beings Woman

Hard Beings Woman or Hurúing Wuhti is central to some Hopi creation stories. She lived before anything else and is the spiritual basis for the matrilineal system of the Hopis. In some versions of the **Hopi creation,** she is the same as **Spider Woman** (see also **Thinking Woman**).

Hawaiian Creation

Polynesians first came to Hawaii in about 750 C.E. Until the nineteenth century the native Polynesian Hawaiians maintained their own religion and their own culture, which would later be dominated by Christian and European-American ways. The Hawaiian creation chant, a 2,000-line poem called the *Kumulipo,* reflects the fertility and lushness of the islands whose creation it celebrates. In the old days the poem was chanted at the births of royal children, signifying a new beginning and the relationship between the child and the plants and animals of the first creation (see also **Polynesian Creation**).

In the beginning there was only the darkness. Out of the darkness, or *ex nihilo*, were born the night and the male, Kumulipo, the essence of darkness, and the female Po'ele, darkness itself (see also **Creation from Nothing**). These were the parents of the children of the darkness: the shellfish of the depths, the plants that grow out of the dark earth, and grubs of the earth. One birth led to another, and then there were many kinds of animals and plants. The world began to grow lighter, but there were no people yet—only the god Kane-i-ka-wai-ola, who watered the plants in the diminishing darkness.

Eventually there was the male Pouliuli, or deep darkness, and the female Powehiwehi, or darkness with a little light. This couple parented the fish of the sea. The fish swam everywhere and multiplied: shark, mackerel, the hilu fish. At the same time things grew and grew on the land, but it was still dark.

Then Po'el'ele, (dark night—male), was born along with Pohaha, (night coming into dawn—female). These were parents to the insects that fly in the night and to the grasshopper, the caterpillar, and the fly. An egg was born, too, and out of that came the bird and then many more birds. It was almost dawn, but really it was still dark.

Now Popanopano and Polalowehi were born, male and female, and they gave birth to the animals that came to the land from the sea—the turtles, lobsters, and geckoes. It was not quite light yet.

Po-kanokano and Po-lalo-uli, male and female, were born at this point, and they began to reproduce. Kamapua'a, the pig, was born. He was dark and beautiful, and his people cultivated the flourishing islands. The footprints of these ancient ones born at the end of night can be seen still.

Po-hiolo the male and Po-ne'a-aku the female, whose names mean night ending, were born. They produced rat, Pilo'i, also near the end of night. The rat people damaged the land with their scratching and eating.

Now came the birth of the male, Po-ne'e-aku, and the female, Poneie-mai, whose names suggest night leaving and night pregnant. They gave birth to dawn, the dog, and the wind, and it was now almost light but still not day.

Po-kinikini and Po-he'enalu gave birth to the time when humans came into the world. Men and women, though different, lived and slept together in deep calmness.

Then finally, dark La'ila'a, the woman, and dark Ki'i, the man, were born. They knew the red-faced god Kane. Now it was daytime and our world.

Source: Leach, 166–171.

Hebrew Creation

For Jews and Christians the Bible is the holy word of God of which Genesis is the first book of five (the Pentateuch or Torah). Genesis contains the creation myth that forms the basis of the Judeo-Christian tradition. Some have seen Genesis as a continuous, uniform story, with Genesis 1:1–2:4a outlining the scheme of the world's origin and Genesis 2:4b–4 carefully painting a more detailed picture of humanity's creation. It appears, however, that the book contains two distinct stories crafted by different hands, strongly influenced by the historical climate experienced by the authors and reminiscent of other ancient Near Eastern stories of creation (see also **Wisdom**).

During the sixth century B.C.E. the Israelite nation faced exile in Babylon. In order to maintain and preserve the beliefs and practices of the monotheistic Israelite tradition, a succession of priestly scholars (P) composed a bold, optimistic, logical hymn portraying an almighty, omnipotent, untouchable god, Elohim, who creates a perfect, beautiful, "good" world populated by creatures made in the likeness of and blessed by God. This is an origin story created for the benefit of a lost nation in the need of encouragement and affirmation, a prelude to the continuing story of a blessed nation. Thus we have Genesis 1:1–2:4a.

> In the beginning God made heaven and earth.
> All was empty, chaotic and dark.
> And God's spirit moved over the watery deep.
> God said, Let light shine and it did.
> And God observed the light, and observed that it was good:
> and God separated the light from the dark . . .

Rather than the high-paced, capricious, ritualistic, magic-filled drama depicted in the Enuma Elish and other Near Eastern creation myths, the

This woodcut from the fifteenth-century Biblia Italiana depicts the six days of the Hebrew creation as told in Genesis, the first book of the Bible's Old Testament.

reader of Genesis 1 is presented with a structured, majestic, logical, somewhat demythologized prehistory. Within the overall seven-day schedule of creation that P describes—a structure that establishes this as a historical text, a precursor to a continuing forward movement of history—another structure emerges in the form of a series of phrases that are repeated before, during, and after acts of creation: "Let there be . . . ," "God said," "And it was so," "God called," "God saw that it was good," "And there was evening and there was morning. . . ." This is a god in total command, creating from thought and *ex nihilo* (see also **Creation by Thought; Creation from Nothing**). In an almost pedantic style, P describes God's creation moving from general terms to specific. The first days comprise a series of separations. Light is created to counter darkness. Time begins. The sky is separated from the earth, and vegetation is separated from the sea (see also **Creation from Division of Primordial Unity; Separation of Heaven and Earth**). On the fourth day the sun and moon are formed to separate the days and nights, creating a calendar. Time is running. On the fifth day living creatures are brought to life on land and in the seas. The culmination of creation is reached when God creates male and female on the sixth day. God blesses them and bids them, "Be fruitful and multiply. . . ." This is a blessing we hear again and again throughout the first five books of the Hebrew Bible. History begins. The seventh day of rest establishes God's ultimate command and power. There is nothing more to be done, to control, or to create because it is all good—creation is perfect in this story.

Unlike the Enuma Elish, this is not a political or cultic treatise. This is a historical text. There are no rituals mentioned. Not even the sabbath is implied in the seventh day of rest. That comes later in the Bible, but, as the Enuma Elish adulates Marduk, Genesis 1:1–2:4a acts as a kind of propaganda for an almighty god, with P's propaganda intentionally countering the polytheistic beliefs of Babylon. Genesis reinforces belief in one all-powerful god. Unlike Marduk of the Enuma Elish, Elohim has ultimate control over the creative process. He is a supreme sovereign without threat of overthrow. Once Elohim creates, it is done, with no chance of change. Marduk is created within the changing cosmos, and one senses that the hierarchy of the Babylonian pantheon could easily shift. Elohim, on the other hand, exists outside of nature. In this creation story he presupposes creation. God is shrouded in mystery. Genesis does not explain where Elohim originates or even the details of how creation occurs. Creation is simply how God commands.

While Genesis 1 reads as a stately, repetitive hymn, there are verses within the text that do not quite fit the style, in that they are so lyrical and almost illogical. Ch. 1:2 reads, "The earth was without form and void, and darkness was upon the face of the waters." The passage leads one to believe that something exists before God begins to create—a sort of primeval darkness perhaps. In Hebrew the word is *Tehom,* which is etymologically related to the name *Tiamat,* the Babylonian primordial sea and goddess of the Enuma Elish. Marduk must conquer and literally carve up the body of Tiamat in order to create the world. While Elohim does not have a physical battle, in this passage, God does create light in order to conquer the chaotic darkness, that darkness "without form and void" (see also **Creation from Chaos**). Myth creeps into P's precise text, acting as a reminder of outside textual influences and as an indirect acknowledgment of the existence of the mysterious darkness and chaos in an otherwise perfect creation.

While the story of Genesis 1 could be considered a demythologized myth, it is hard to ignore the influences of other myths upon it. One could at least say this story was written in reaction to creation myths of nearby cultures; for instance, Genesis includes mention of great sea monsters who are reduced in status to ordinary living creatures, a theme that hearkens back to Tiamat, who is often referred to as a monster and primordial sea. In other Near Eastern mythologies, the sun and moon are gods who have names and rule. P tells of their creation on the fourth day as simply luninaries without name or function except to keep time.

Verses 26 and 27 describe God as making man in "our image . . . in the image of God he created him; male and female he created them." The plural reference to God, according to some, possibly suggests the presence of a divine council of some sort that hearkens back to a polytheistic conception of the cosmos. Unlike the Enuma Elish, in which men are created to serve the gods, men in Genesis are made masters who have dominion over creation. While in the Enuma Elish the earth and its inhabitants are created almost haphazardly, as needed, Elohim creates with an unalterable plan in mind.

During approximately the tenth century B.C.E. Israel was a strong nation, a people with visions of divine power. During the reign of Solomon or shortly thereafter, there developed another account of creation, written by a poet-storyteller (J) who speaks of a not-so-perfect creation, one in which humans are made from dust. The creator is an anthropomorphized god who provides temptations and inflicts punishment on those with visions of grandeur—those who might challenge his power. While P describes the sweeping creation of our world, J speaks of a series of damaged relationships between people and the environment, between man and woman, and most importantly between humanity and God.

Although J's Genesis (2–4) describes a fantastical drama complete with a tree of life, a tree of knowledge, a talking serpent, a blissful lack of shame between man and woman, a paradisiacal garden, and a supreme sovereign who walks among the trees speaking with the creatures he has formed from the dust, the story is more real to the audience than the mysterious, inaccessible, logical, demythologized report of Genesis 1–2:4.

The story told by J, who is a poet and theologian rather than a priest, provides a setting, the Garden of Eden—a place certainly well outside human experience. Nevertheless J describes a drama that presents issues ubiquitous to human experience: temptation, choice and the consequences of choice, relationships, compliance, tension, blame, and punishment. Genesis 2–4 is a human interest story, while Genesis 1 is a piece of royal propaganda.

Unlike Genesis 1, which describes a watery chaos and the submission of chaos to order, thus resembling Babylonian creation myths, Genesis 2 describes the beginning as a piece of barren earth that bears life only when a mist comes to water the earth, thus displaying a Canaanite influence. In Babylonia, floods were a yearly threat, but to the desert-dwelling, nomadic Canaanites, water was always welcome and meant new life. In this light the Garden of Eden becomes an interesting image. It is an appropriate metaphor, located at the center of four rivers, umbilical cords of life in a Canaanite tradition. It is humanity's womb, a place without outside threat where one can never return, the birthplace of humanity's relationship with God. Like Genesis 1, Genesis 2 is a prehistory.

While P stresses the inaccessability of God and the clear separation between God and his creations, J attempts to show the Lord God (Yahweh) as somewhat of an experimenter. In a very small sense J's God reminds one of Marduk in that he creates as he sees need—for instance the animals and then woman.

As in the Enuma Elish, humans in Genesis are created from clay, and man works for God. He tends the garden and names the plants and animals, but unlike in the Enuma Elish, God creates a paradise specifically for man, has a relationship with him, and treats him as a kind of god. Adam, it should be noted, is not a proper name at this point. It simply means humanity and is also a possible play on the Hebrew word for earth, which is *adamah* (see also **Adam**). Adam's mate is named **Eve,** which means life. These names are appropriate, and in using them J stresses the importance of the environment/earth and humanity's relationship to it.

It has often been assumed that the creation of woman from the rib of man in J's text makes her the inferior of the pair. If one reads the text, however, God admits that man is not complete and needs a fit helper. He tries to find this helper by creating animals. This experiment allows J to indicate his conception of the proper relationship between nature and humans: we should be close but not companions. It is not until God creates a woman that humanity is complete. Man cries out in fact that woman is flesh of his flesh. J adds that "they become one flesh." At last creation is complete and seemingly perfect up to this point.

There is a text in the apocryphal tradition that mentions another first woman, Lilith. Lilith apparently wanted to be sexually dominant. The patriarchy did not consider this a proper version of the story; Lilith retreated into obscurity, and Eve became Adam's companion or helpmate. Perhaps a remnant of Lilith remains in Eve; however, she is the first to try the fruit from the Tree of Knowledge. This act has often been interpreted as an indication of the weakness of an inferior sex. One might also see it as the act of a more independent human. It seems that neither interpretation is necessarily the right one, since up to this point Adam and Eve have been portrayed as equal partners, as literally "humanity" and "life." Why would J upset this balance? This story is not really about male and female roles.

The serpent in this myth represents temptation rather than evil. In the ancient Near East the serpent had an ambiguous role. The snake merely helps Eve come to her own decision, urges her to do something she already has in mind. There is temptation because God gives us choices from the beginning. Adam and Eve choose not to obey God's commandment. The story of Genesis 3 is one about disobeying a commandment rather than a story about what the knowledge of good and evil is. J presents issues to his readers that are pertinent. Obedience and choice are issues with which we still contend. J focuses on neither good nor evil in connection with the eating of the fruit, but rather the shame and fear. Adam tells God that he heard his (God's) voice and was afraid—because he was naked.

Tension between man and woman begins here. Adam blames Eve for offering him the forbidden fruit. Tensions between man and nature also begin. Eve blames the serpent, and Eve's children and the snake's children are forever enemies. Ultimately man is expelled from the paradise and forced to work the land to survive. Adam and Eve both disobey God's commandment. God gives Adam and Eve the choice to decide what kind of relationsip they will have with him. People often interpret this commandment not to eat the fruit of the Tree of Knowledge as an act of a cruel God who gives us a test he knows we will fail, or as a way of withholding information of good and evil—information that could potentially threaten God's power. It makes more sense to interpret the commandment as God giving us freedom to choose. If he had not given that choice, our relationship with him would always have been one of absolute dependence.

It is interesting that God's prohibition is against eating. This is seemingly trivial, but eating meals and prohibitions against certain foods are part of great commandments later in the Bible.

The punishment for Adam and Eve's choice is death, not literal perhaps—they continue to live for 900 years or so—but figurative. They are no longer immortal and have destroyed a paradisiacal relationship with God. They are thrown out of the Garden of Eden and must struggle to survive and learn what pain means. J presents life as we know it, one of choices and hardship.

What follows is the familiar King James translation of a part of the Hebrew creation.

From the First Book of Moses Called Genesis

Chapter 1

In the beginning God created the heaven and the earth.

[2] And the earth was without form, and void; and darkness *was* upon the face of the deep. And the Spirit of God moved upon the face of the waters.

[3] And God said, Let there be light: and there was light.

[4] And God saw the light, that *it was* good: and God divided the light from the darkness.

[5] And God called the light Day, and the darkness he called Night. And the evening and the morning were the first day.

[6] And God said, Let there be a firmament in the midst of the waters, and let it divide the waters from the waters.

[7] And God made the firmament, and divided the waters which *were* under the firmament from the waters which *were* above the firmament: and it was so.

[8] And God called the firmament Heaven. And the evening and the morning were the second day.

[9] And God said, Let the waters under the heaven be gathered together unto one place, and let the dry *land* appear: and it was so.

[10] And God called the dry *land* Earth; and the gathering together of the waters called he Seas: and God saw that *it was* good.

[11] And God said, Let the earth bring forth grass, the herb yielding seed, *and* the fruit tree yielding fruit after his kind, whose seed *is* in itself, upon the earth: and it was so.

[12] And the earth brought forth grass, *and* herb yielding seed after his kind, and the tree yielding fruit, whose seed *was* in itself, after his kind: and God saw that *it was* good.

[13] And the evening and the morning were the third day.

[14] And God said, Let there be lights in the firmament of the heaven to divide the day from the night; and let them be for signs, and for seasons, and for days, and years:

[15] And let them be for lights in the firmament of the heaven to give light upon the earth: and it was so.

[16] And God made two great lights; the greater light to rule the day, and the lesser light to rule the night: *he made* the stars also.

[17] And God set them in the firmament of the heaven to give light upon the earth,

[18] And to rule over the day and over the night, and to divide the light from the darkness: and God saw that *it was* good.

[19] And the evening and the morning were the fourth day.

[20] And God said, Let the waters bring forth abundantly the moving creature that hath life, and fowl *that* may fly above the earth in the open firmament of heaven.

[21] And God created great whales, and every living creature that moveth, which the waters brought forth abundantly, after their kind, and every winged fowl after his kind: and God saw that *it was* good.

[22] And God blessed them, saying, Be fruitful, and multiply, and fill the waters in the seas, and let fowl multiply in the earth.

[23] And the evening and the morning were the fifth day.

[24] And God said, Let the earth bring forth the living creature after his kind, cattle, and creeping thing, and beast of the earth after his kind: and it was so.

[25] And God made the beast of the earth after his kind, and cattle after their kind, and every thing that creepeth upon the earth after his kind: and God saw that *it was* good.

[26] And God said, Let us make man in our image, after our likeness: and let them have dominion over the fish of the sea, and over the fowl of the air, and over the cattle, and over all the earth, and over every creeping thing that creepeth upon the earth.

[27] So God created man in his *own* image, in the image of God created he him; male and female created he them.

[28] And God blessed them, and god said unto them, Be fruitful, and multiply, and replenish the earth, and subdue it: and have dominion over the fish of the sea, and over the fowl of the air, and over every living thing that moveth upon the earth.

There are further references to creation in the Bible that should be mentioned. In Psalms 33:6–15, we are told that the heavens were created "by the Word of the Lord" (see also **Creation by Word**) and "the breath of his mouth." Furthermore, "He gathereth the waters of the sea together as an heap." Psalm 104, too, reveals God as an active creator "who laid the foundations of the earth." In the book of Job, God reveals his role in creation with great clarity:

Chapter 38

Then the Lord answered Job out of the whirlwind, and said,

[2] Who *is* this that darkeneth counsel by words without knowledge?

[3] Gird up now thy loins like a man for I will demand of thee, and answer thou me.

[4] Where wast thou when I laid the foundations of the earth? declare, if thou hast understanding.

[5] Who hath laid the measures thereof, if thou knowest? or who hath stretched the line upon it?

[6] Whereupon are the foundations thereof fastened? or who laid the corner stone thereof;

[7] When the morning stars sang together, and all the sons of God shouted for joy?

[8] Or *who* shut up the sea with doors, when it brake forth, *as if* it had issued out of the womb?

[9] When I made the cloud the garment thereof, and thick darkness a swaddlingband for it,

[10] And brake up for it my decreed place, and set bars and doors,

[11] And said, Hitherto shalt thou come, but no further: and here shall thy proud waves be stayed?

[12] Hast thou commanded the morning since thy days; *and* caused the dayspring to know his place;

[13] That it might take hold of the ends of the earth, that the wicked might be shaken out of it?

[14] It is turned as clay *to* the seal; and they stand as a garment.

15 And from the wicked their light is withholden, and the high arm shall be broken.

16 Hast thou entered into the springs of the sea? or hast thou walked in the search of the depth?

17 Have the gates of death been opened unto thee? or hast thou seen the doors of the shadow of death?

18 Hast thou perceived the breadth of the earth? declare if thou knowest it all.

19 Where *is* the way *where* light dwelleth? and *as for* darkness, where *is* the place thereof,

20 That thou shouldest take it to the bound thereof, and that thou shouldest know the paths *to* the house thereof?

21 Knowest thou *it,* because thou wast then born? or *because* the number of thy days *is* great?

22 Hast thou entered into the treasures of the snow? or hast thou seen the treasures of the hail,

23 Which I have reserved against the time of trouble, against the day of battle and war?

24 By what way is the light parted, *which* scattereth the east wind upon the earth?

25 Who hath divided a watercourse for the overflowing of waters, or a way for the lightning of thunder;

26 To cause it to rain on the earth, *where* no man *is; on* the wilderness, wherein *there is* no man;

27 To satisfy the desolate and waste *ground;* and to cause the bud of the tender herb to spring forth?

28 Hath the rain a father? or who hath begotten the drops of dew?

29 Out of whose womb came the ice? and the hoary frost of heaven, who hath gendered it?

30 The waters are hid as *with* a stone, and the face of the deep is frozen.

31 Canst thou bind the sweet influences of Ple'iades, or loose the bands of O-ri'on?

32 Canst thou bring forth Maz'za-roth in his season? or canst thou guide Arc-tu'rus with his sons?

33 Knowest thou the ordinances of heaven? canst thou set the dominion thereof in the earth?

34 Canst thou lift up thy voice to the clouds, that abundance of waters may cover thee?

[35] Canst thou send lightnings, that they may go, and say unto thee, Here we *are?*

[36] Who hath put wisdom in the inward parts? or who hath given understanding to the heart?

[37] Who can number the clouds in wisdom? or who can stay the bottles of heaven,

[38] When the dust groweth into hardness, and the clods cleave fast together?

[39] Wilt thou hunt the prey for the lion? or fill the appetite of the young lions,

[40] When they couch in *their* dens, *and* abide in the covert to lie in wait?

[41] Who provideth for the raven his food? when his young ones cry unto God, they wander for lack of meat.

Sources: Bible, Genesis 1–3; Gowan, 34; O'Brien and Major.

Hesiod

Hesiod was the eighth-century Greek poet responsible for writing down the best known of the **Greek creation** myths in his *Theogony* and *Works and Days.*

Hindu Creation

See Indian Creation.

Hopi Creation

As is appropriate for a society that is to this day matrilineal (that is, a person belongs primarily to the mother's family and property is passed down from the mother), the Hopi creation myth is dominated by the female creative principle, **Hard Beings Woman** (Hurúing Wuhti) or sometimes **Spider Woman** or both. Like the female element in other creation myths, this woman is associated with the **Earth Mother**, and, not surprisingly, the creation is of the emergence type (see also **Creation by Emergence**). The male principle, Tawa, is the divine creative energy represented by the sun. Familiar motifs in the Hopi myth include the division of the divine parents into new forms, **creation by thought,** and the idea of creation by song, which perhaps has its source in the ritual song-dance ceremonies the Hopis learned from their ancient ancestors, the Anasazis (see also **Creation by Word; Creation from Division of Primordial Unity**).

One version of the Hopi creation says that in the beginning were Spider Woman (the earth goddess) and Tawa (the sun god). Tawa controlled heaven and its mysteries; Spider Woman's precinct was earth. The two lived in Spider Woman's lair, Under-the-World. They were all-in-all in those days but were not husband and wife.

Longing for company, Tawa divided himself into Tawa and Muiyinwuh, the god of life energy. Spider Woman divided into Spider Woman and Huzruiwuhti, goddess of life forms. Huzruiwuhti and Tawa became a couple, and from their union came the Holy Twins, the Four Corners, the Up and the Down, and the Great Serpent.

Spider Woman and Tawa had the Sacred Thought, which was of placing the world between the Up and Down within the Four Corners in the void that was the Eternal Waters. The Thought became the first song: "Father of all, Life and Light I am," sang Tawa. "Mother of all, receiver of Light and weaver of Life I am," sang Spider Woman.

"My Thought is of creatures that fly in the Up and run in the Down and swim in the Eternal Waters," sang the god.

"May the Thought live," intoned Spider Woman, and she formed it of clay.

Together the first gods placed a sacred blanket over the new beings and chanted the song of life. The beings stirred into Life.

The first gods were not yet pleased, however, so Tawa thought of beings like Spider Woman and himself, who could rule over the world between the Up and Down within the Four Corners, and Spider Woman formed the new thought into man and woman, whom she cradled in her arms until they breathed life.

Instructed to multiply, the man, woman, and other creatures did so, and when there were enough, Spider Woman led them through the four worlds of Under-the-World to the spider hole or *sipapu* that led to the world between Up and Down within the Four Corners, where Tawa sent his all-seeing warmth and light as he made a daily journey overhead.

Spider Woman made clans, established villages, and instructed the people on how to keep life and form. The men must hunt, she said, and grow the corn. The women must build and keep the houses and lead the families. The men must build underground places—kivas—where clans would meet and send messages to the Gods and from which the people could emerge for new beginnings after hearing the myth of their original emergence into the world. She told the people how to call the Great Serpent, who would strike the earth and give rain to make the crops grow. Before descending back to Under-the-World, Spider Woman called all the clans together. "Do as I have taught," she said, "and you will prosper. Be sure that Tawa and I will always watch over you."

In some Hopi myths it is Tawa or Taiowa who is the original creator.

Most often it is Hard Beings Woman, Huruing Wuhti, who at the beginning was the only hard surface—the nucleus of the world to come. Sun existed too. He came from the east each night to Hard Beings' kiva in the west, and she let him in so he could take the trap door to the underworld to find his way back east. Huruing Wuhti also created Muingwu, the young crop god (see also **Acoma Creation**).

The following story exemplifies a Hopi emergence myth as told to Henry Voth in 1905.

> A very long time ago there was nothing but water. In the east Hurúing Wuhti, the deity of all hard substances, lived in the ocean. Her house was a kiva like the kivas of the Hopi of to-day. To the ladder leading into the kiva were usually tied a skin of gray fox and one of a yellow fox. Another Hurúing Wuhti lived in the ocean in the west in a similar kiva, but to her ladder was attached a turtle-shell rattle.
>
> The Sun also existed at that time. Shortly before rising in the east the Sun would dress up in the skin of the gray fox, whereupon it would begin to dawn—the so-called white dawn of the Hopi. After a little while the Sun would lay off the gray skin and put on the yellow fox skin, whereupon the bright dawn of the morning—the so-called yellow dawn of the Hopi—would appear. The Sun would then rise, that is, emerge from an opening in the north end of the kiva in which Hurúing Wuhti lived. When arriving in the west again, the sun would first announce his arrival by fastening the rattle on the point of the ladder beam, whereupon he would enter the kiva, pass through an opening in the north end of the kiva, and continue his course eastward under the water and so on.
>
> By and by these two deities caused some dry land to appear in the midst of the water, the waters receding eastward and westward. The Sun passing over this dry land constantly took notice of the fact that no living being of any kind could be seen anywhere, and mentioned this fact to the

two deities. So one time the Hurúing Wuhti of the west sent word through the Sun to the Hurúing Wuhti in the east to come over to her as she wanted to talk over this matter. The Hurúing Wuhti of the east complied with this request and proceeded to the west over a rainbow. After consulting each other on this point the two concluded that they would create a little bird; so the deity of the east made a wren of clay, and covered it up with a piece of native cloth (möchápu). Hereupon they sang a song over it, and after a little while the little bird showed signs of life. Uncovering it, a live bird came forth, saying: "uma hínok pas nui kita náwakna?" (why do you want me so quickly). "Yes," they said, "we want you to fly all over this dry place and see whether you can find anything living." They thought that as the Sun always passed over the middle of the earth, he might have failed to notice any living beings that might exist in the north or the south. So the little Wren flew all over the earth, but upon its return reported that no living being existed anywhere. Tradition says, however, that by this time Spider Woman (Kóhkang Wuhti), lived somewhere in the south-west at the edge of the water, also in a kiva, but this the little bird had failed to notice.

Hereupon the deity of the west proceeded to make very many birds of different kinds and form, placing them again under the same cover under which the Wren had been brought to life. They again sang a song over them. Presently the birds began to move under the cover. The goddess removed the cover and found under it all kinds of birds and fowls. "Why do you want us so quickly?" the latter asked. "Yes, we want you to inhabit this world." Hereupon the two deities taught every kind of bird the sound that it should make, and then the birds scattered out in all directions.

Hereupon the Hurúing Wuhti of the west made of clay all different kinds of animals, and they were brought to life in the same manner as the birds. They also asked the same question: "Why do you want us so quickly?" "We want you to inhabit this earth," was the reply given them, whereupon they were taught by their creators their different sounds or languages, after which they proceeded forth to inhabit the different parts of the earth. They now concluded that they would create man. The deity of the east made of clay first a woman and then a man, who were brought to life in exactly the same manner as the birds and animals before them. They asked the same question, and were told that they should live upon this earth and should understand everything. Hereupon the Hurúing Wuhti of the east made two tablets of some hard substance, whether stone or clay tradition does not say, and drew upon them with the wooden stick certain characters, handing these tablets to the newly created man and woman, who looked at them, but did not know what they meant. So the deity of the east rubbed with the palms of her hands, first the palms of the woman and then the palms of the man, by which they were enlightened so that they understood the writing on the tablets. Hereupon the deities taught these two a language. After they had taught them the language, the goddess of the east took them out of the kiva and led them over a rainbow, to her home in the east. Here they stayed four days, after which Hurúing Wuhti told them to go now and select for themselves a place and live there. The two proceeded forth saying that they would travel around a while and wherever they would find a good field they would remain. Finding a nice place at last, they built a small, simple house, similar to the old houses of the Hopi. Soon the Hurúing Wuhti of the west began to think of the matter again, and said to herself: "This is not the way yet that it should be. We are not yet done," and communicated her thoughts to the Hurúing Wuhti of the east. By this time Spider Woman had heard about all this matter and she concluded to anticipate the others and also create some beings. So she also made a man and woman of clay, covered them up, sang over them, and brought to life her handiwork. But these two proved to be Spaniards. She taught them the Spanish language, also giving them similar tablets and imparting knowledge to them by rubbing their hands in the same manner as the woman of the East had done with the "White Men." Hereupon she created two burros, which she gave to the Spanish man and woman. The latter settled down close by. After this, Spider Woman continued to create people in the same manner as she had created the Spaniards, always a man and a woman, giving a different language to each pair. But all at once she found that she had forgotten to create a woman for a certain man, and that is the reason why now there are always some single men.

She continued the creating of people in the same manner, giving new languages as the pairs formed. All at once she found that she had failed to create a man for a certain woman, in other words, it was found that there was one more woman than there were men. "Oh my!" she said, "How is this?" and then addressing the single woman she said: "There is a single man somewhere, who went away from here. You try to find him and if he accepts you, you live with him. If not, both of you will have to remain single. You do the best you can about that." The two finally found each other, and the woman said, "Where shall we live?" The man answered: "Why here, anywhere. We shall remain together." So he went to work and built a house for them in which they lived. But it did not take very long before they commenced to quarrel with each other. "I want to live here alone," the woman said. "I can prepare food for myself." "Yes, but who will get the wood for you? Who will work the fields?" the man said. "We had better remain together." They made up with each other, but peace did not last. They soon quarreled again, separated for a while, came together again, separated again, and so on. Had these people not lived in that way, all the other Hopi would now live in peace, but others learned it from them, and that is the reason why there are so many contentions between the men and their wives. These were the kind of people that Spider Woman had created. The Hurúing Wuhti of the west heard about this and commenced to meditate upon it. Soon she called the goddess from the east to come over again, which the latter did. "I do not want to live here alone," the deity of the west said, "I also want some good people to live here." So she also created a number of other people, but always a man and a wife. They were created in the same manner as the deity of the east had created hers. They lived in the west. Only wherever the people that Spider Woman had created came in contact with these good people there was trouble. The people at that time led a nomadic life, living mostly on game. Wherever they found rabbits or antelope or deer they would kill the game and eat it. This led to a good many contentions among the people. Finally the

Woman of the west said to her people: "You remain here; I am going to live, after this, in the midst of the ocean in the west. When you want anything from me, you pray to me there." Her people regretted this very much, but she left them. The Hurúing Wuhti of the east did exactly the same thing, and that is the reason why at the present day the places where these two live are never seen.

Those Hopi who now want something from them deposit their prayer offerings in the village. When they say their wishes and prayers they think of those two who live in the far distance, but of whom the Hopi believe that they still remember them.

The Spanish were angry at Hurúing Wuhti and two of them took their guns and proceeded to the abiding place of the deity. The Spaniards are very skillful and they found a way to get there. When they arrived at the house of Hurúing Wuhti the latter at once surmised what their intentions were. "You have come to kill me," she said; "don't do that; lay down your weapons and I shall show you something; I am not going to hurt you." They laid down their arms, whereupon she went to the rear end of the kiva and brought out a white lump like a stone and laid it before the two men, asking them to lift it up. One tried it, but could not lift it up, and what was worse, his hands adhered to the stone. The other man tried to assist him, but his hands also adhered to the stone, and thus they were both prisoners. Hereupon Hurúing Wuhti took the two guns and said: "These do not amount to anything," and then rubbed them between her hands to powder. She then said to them: "You people ought to live in peace with one another. You people of Spider Woman know many things, and the people whom we have made also know many, but different, things. You ought not to quarrel about these things, but learn from one another; if one has or knows a good thing he should exchange it with others for other good things that they know and have. If you will agree to this I shall release you. They said they did, and that they would no more try to kill the deity. Then the latter went to the rear end of the kiva where she disappeared through an opening in the floor, from where she exerted a secret influence upon the stone and thus released the two men. They departed, but Hurúing

Wuhti did not fully trust them, thinking that they would return, but they never did.

Reprinted from H. R. Voth, *The Traditions of the Hopi,* Anthropological Series, vol. 8, Chicago: Field Columbian Museum, 1905.

Sources: Erdoes, 115–117; Mullett, 1–7; Sproul, 268–284; Tyler, 82–84; Williamson, 62–65.

Hottentot Creation

The Hottentots, originally from the Cape of Good Hope, now live mostly in Namibia. Their language, including the "click" sound, and their religion resemble those of the Bushmen (see also **Bushmen Creation**). This creation myth contains aspects of the **world parent creation** myth.

The Hottentot people come from Tsui-[click]-Goab, which once fought a long and difficult battle against a negative force, [click]-Gaunab. As the evil one was dying he wounded the good one's knee, leading to the name Tsui-[click]-Goab, which means sore knee. The lame god is the provider of all good things: rain, crops, animals. His home is in heaven. The Evil One lives in another, dark heaven.
Source: Sproul, 34–35.

Hungarian (Magyar) Creation

See Finnish Creation.

Huron Creation

The Huron Indians have always been an oppressed people. Facing many defeats over the centuries, they have migrated from place to place—from Ontario to New York, Ohio, Kansas, and Oklahoma. Familiar figures in their creation myth, which is of the earth-diver variety (see also **Earth-Diver Creation**) are the woman from the sky and the twins (see also **Cherokee Creation**).

In the beginning there was water and water animals. Into this emptiness a goddess fell out of the sky. Two loons saw her falling, and they made themselves into a cushion for her to land on. The loons cried out for help from the other animals; the loon still has a loud voice. It was Giant Tortoise who came first and offered to carry Sky Woman on his back. Then he called a council to see what could be done. It was decided that Sky Woman must have some place permanent to live, so Tortoise ordered the animals to dive into the water to find earth. Many died in their attempt. Only the toad made it and returned with a pinch of earth. Sky Woman took it, placed it on Tortoise's back, and it grew into land. The Great Tortoise still supports the world.

Sky Woman had been pregnant when she fell. She was carrying twin boys—one good, one bad. They fought within her womb. The bad one refused to be born properly and killed his mother by breaking through her side. From the buried goddess came all of the vegetables and fruits of the earth. From then on life was marked by the struggle between the good and bad brothers. The good one made streams to travel on, the bad one made rapids to impede travel on streams. One made useful animals, the other dangerous ones, and so forth. Eventually the good brother killed the bad brother and prepared a place for the good dead in the west.

Source: Long, 193–196.

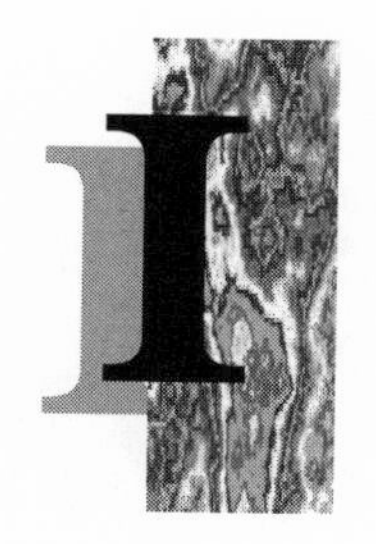

Icelandic (Norse) Creation

The primary source for the creation myths of the Germanic peoples of Scandinavia and Iceland (the Norse people, sometimes called Vikings) is contained in the Icelandic text called the Younger Edda or the Prose Edda, compiled by the Icelandic historian Snorri Sturluson (1173–1241) in about 1220 C.E. Snorri's work is based on much older works from the oral tradition and from the Elder Edda or Poetic Edda. The creation myth of the first part of the Prose Edda concerns the Ice Giant, Ymir or Imir, from whose body the world was made. Note the connection with the **Celtic creation** myth.

Long ago King Gylfi ruled what is now Sweden. He learned from a wise old woman about the Aesir, the gods who live in Asgard, or Valhalla. Gylfi disguised himself as an old man, decided to call himself Gangleri, and made his way to Valhalla. There he met the High One, who answered his various questions about the world and its origins.

The High One told Gangleri that once there were two places, one in the south that was all fire and light and one in the north that was icy and dark. The first was called Muspell and the second Niflheim. The two atmospheres met in an emptiness between them called Ginnungagap. There the hot and the cold mixed and caused moisture to form and life to begin, first as the evil frost giant Ymir.

Ymir lay down in Ginnungagap and gave birth to a man and a woman from his armpits; one of his legs mated with the other to make a son. Thus began the family of frost ogres. Some of the melting ice became the cow giant, Auohumla, whose teats flowed with rivers of milk to feed the giant and his family.

Norse myths attributed creation to the giant Ymir who was nourished by Audhumla, a sacred cow. In an eighteenth-century painting Audhumla frees the first man, Buri the Strong, from ice, as the young giant Ymir feeds from her.

As for the cow, she fed on the ice blocks around her. As she licked the ice, a man gradually appeared from it. He was Buri the Strong; he had a son called Bor who married Bestla, a daughter of one of the frost ogres. Bor and Bestla produced the great god Odin and the gods Vili and Ve. These gods killed Ymir, and from the blood that resulted, all the frost ogres were destroyed in a flood. One giant, Bergelmir, escaped with his wife and family.

The three gods took Ymir's body to the center of Ginnungagap and turned his body into the earth and his blood into the seas. His bones became the mountains and his teeth and jaws became rocks, stones, and pebbles. The gods turned his skull into the sky, held up at each of the four corners by a dwarf. These are called by the names of the four directions. From Muspell they took sparks and embers and made the sun, moon, and stars, placing them over Ginnungagap (see also **Animism**).

The earth was made to be round and surrounded by the ocean. The gods gave shore lands to the descendants of the surviving giant family. They made a stronghold out of Ymir's eyebrows, and they made clouds out of his brains.

The three gods made man and woman out of two fallen trees, an ash and an elm. Odin breathed life into the new pair. Vili's gift to them was intelligence, and Ve's gifts were sight and hearing. The first man was named Ask; the first woman was Embla. The stronghold, Midgard, became their home, so they were protected from the cruel giants outside.

The following lines are in the Prose Edda as remembered from the Poetic Edda:

From Ymir's flesh
the earth was made
and from his blood the seas,
crags from his bones,
trees from his hair,
and from his skull the sky.

From his eyebrows
the blessed gods
made Midgard for the sons of men,
and from his brains
were created
all storm-threatening clouds.

Source: Sturluson, 36, The Prose Edda.

Ijaw Creation

This **creation from chaos** story from Nigeria is a small part of an epic story of the heroine Ogboinba, who oversteps her proper boundaries in an attempt to force the creatrix—Woyengi—to overcome her barrenness by recreating her. As a punishment for her arrogance, Ogboinba is condemned to live forever in the eyes of pregnant women and other people, too. When you look straight into somebody's eyes, say the Ijaw, Ogboinba looks back at you.

Once upon a time there was a field containing a gigantic iroko tree (see also **Yggdrasil**). Into that field suddenly descended a table with a great pile of dirt on it, a chair, and a very large "creation stone." Then, announced by thunder and lightning, Woyengi came down, sat on the chair, rested her feet on the sacred stone, took the earth from the table, and made the first humans out of it (see also **Creation from Clay**).

These first humans needed life, so Woyengi held them and breathed breath into them, and they gained life. They were neither male nor female, so she ordered each to choose a gender, and they did. Now she asked the new men and women each to choose a way of life and a way of death. So it was that those who chose riches got riches, those who chose children got children, those who chose to die from smallpox got smallpox, and so forth. All types of lives and all types of diseases and other death-bringing activities were chosen on that day.

One woman asked for successful children, and Ogboinba asked for sacred powers. These women chose to be born in the same village. In fact, they were even born of the same parents, and they lived together as loving sisters.

When it was time, the girls took husbands; the first sister produced many wonderful children, but Ogboinba produced only magic. She became unhappy and went off to find Woyengi to have her way of life changed. This was a mistake, because her way of life had already been chosen. That is why eventually she angered the goddess and had to hide in people's eyes.

Source: Beier, 23–41.

Imperfect Creation

Sometimes the creator is dissatisfied with his work and decides to leave or destroy it and begin over again. Sometimes an element of creation fails. The clay, for instance, might be bad clay, reflecting the problems that potters and other craftspeople often have with their materials. In the

Quiché Mayan creation, for example, the clay is bad, and the creator has to start over. Sometimes the creator realizes after the fact that his creation—especially the human aspect of it—is wrong, and he sends a cleansing flood (see also **The Fall of Humankind; The Flood and Flood Hero**).

Inca Creation

The Incas, whose great Peruvian-based civilization was conquered by Pizarro in 1553, had a highly developed society that was in some ways more advanced than that of its conqueror. The Incas had several versions of a **creation from chaos** myth that resembles the walkabouts or the Dreaming of the Australian aborigines (see also **The Dreaming**).

The pre-Incas worshipped the sun, Pachacamac. Later, in Incan times, he was sometimes called Viracocha (Wiraqoca). He was beyond comprehension, unnamed, undefinable. The ancient people had simply taught that he rose in the beginning out of Lake Titicaca, and that he had made the stars, planets, and moon and sent them on their way.

There were Incas who said Pachacamac created everything, including humans, out of clay. Out of pity he sent his son and daughter, born of the moon goddess, down to teach the miserable people how to plant food, make houses, and weave clothes. His children lived at Lake Titicaca, but they were free to come and go as they chose as long as at every stop they stuck into the ground a golden rod their father gave them. Each mark of the rod would be a sign to the people to build a city. "Teach the people to be kind and good," commanded Pachacamac; "I will provide warmth and light." Pachacamac's son was known as the Inca (the emperor) and his sister was his queen. All Inca rulers descended from this first pair.

The Inca and his sister-wife stopped in the valley of Huanacauri and succeeded in planting the golden rod; in fact, it sunk into the earth and disappeared. This is why there is a temple to the sun there to this day. The Inca took this as a sign that Pachacamac wished him to settle people there. He and his sister went out separately into the surrounding country to find the people. The Inca went north and his queen went south, preaching Pachacamac's rule. They wore wonderful clothes and ornaments, and they had elaborately pierced ears. The people were moved by them, and they believed and did as they were told. So it was that they all gathered at the place that would become the capital city. The Inca and his followers founded Northern Cuzco (Hanan-Cuzco), and the queen and her converts founded Southern Cuzco (Hurin-Cuzco). All Incan cities and villages forever after divided into upper and lower halves, representing male and female and all of the useful opposites.

In those early days, the Inca taught men their tasks, farming and building. The queen taught the women to do the things they still do—weaving and cooking, for instance. Over the years, the first Inca developed an army to bring all the people around into his benevolent care.

Another version of this story holds that there were three caves on a hill. They were called Tavern of the Dawn (Paccari-tambo). Out of the central cave emerged the four brothers, Ayar-manco (the leader), Ayar-auca (the warrior), Ayar-cachi (the salty one), and Ayar-oco (the peppery one), as well as their sister-wives, Mamaocclo (the pure one), Mama-huaco (the fighter), Mama-ipagora, and Mama Rawa (see also **Incest in Creation Myths**).

Led by Ayar-manco and his golden rod, the brothers and their sister-wives traveled about inventing the world. Ayar-cachi was dangerously strong, however, so his companions tricked him into a cave and rolled a rock in front of it to lock him in. This is the Traitor's Stone, which is not far from Cuzco.

Near the valley of Cuzco the group saw a sacred idol, and when Ayar-oco touched it, he turned to stone, the stone that is now Huanacauri on Huanacauri Hill.

When the brothers and sisters rested, Ayar-manco thrust the golden rod into the ground, and it sunk deep into the earth. This was a clear sign, so Ayar-manco ordered Ayar-auca to stand on a cairn there to proclaim the new site for a city. At this point Ayar-auca turned into a stone, the cornerstone for the city to be.

Ayar-Manco built Cuzco, the capital of the Incas, high in the mountains. He and Mama-occlo became parents of the first Inca, Sinchi Roq'a. Ayar-Manco also built the House of the Sun, the great temple to the sun-creator, Pachacamac (Viracocha).

Sources: Leach, 115–118; Olcott, 26; Sproul, 301–305.

Incest in Creation Myths

As the first people are children of the same creator, incest is necessarily an aspect of creation myths. In some cases, however, such as the Egyptian and Incan creations, it seems to have a symbolic import, suggesting the strength of dynasties or of cultural origins (see also **Greek Creation; Hebrew Creation; Maori Creation; Miwok Creation; Polynesian Creation**). In a few cases, such as in the Egyptian creation, the brother-sister incest motif was reinforced in actual social practice; that is, pharaohs married their sisters to keep the dynasty pure. Usually, though, the presence of the incest motif in creation myths does nothing to undermine the incest taboo in social practice. The motif is symbolic rather than representative.

Indian Creation

There are many expressions of creation in the Hindu tradition of India. This variety mirrors the variety of gods, and we must remind ourselves that all of the creations, like all of the gods, are in reality metaphors for a single absolute principle eventually called **Brahman,** a principle that is everywhere and nowhere, everything and nothing. Creation in Hindu mythology can be seen as originating *ex nihilo* in Brahman's thought or can be attributed to the actions of the god Brahma, the personified masculine expression of Brahman (see also **Creation from Nothing**). It may also be the work of Prajapati, the Vedic progenitor who, like the primal man, Purusa (another version of the progenitor), merges into the figure of Brahma. As the creation myths below suggest, Hindu mythology—which is to say Hindu religion—is among the richest and most sophisticated the world has produced.

Hindu creation stories come from the Hindu scriptures and epics. The oldest creation myths are found in the **Rig Veda,** a collection of hymns. This is the first of the four Vedas composed by the Aryan invaders (Indo-Europeans), who came into India from Iran in the second millennium B.C.E. The Vedas form the basis for Vedism, the early form of Hinduism, in which an apparent polytheism is in effect.

An early myth emerges in fragments from the first and tenth books of hymns in the Rig Veda. In the myth are the familiar themes of the sacrificial basis of creation (see also **Babylonian Creation**), incest (see also **Incest in Creation Myths**), and the separation of the first "parents," who, although of opposite gender, resemble the traditional depiction of the Egyptian Nut (heaven) arched over Geb (earth). We are also reminded of the Greek story of the merging of heaven and earth as Uranos and Gaia (see also **Creation from Division of Primordial Unity; Egyptian Creation; Greek Creation; Separation of Heaven and Earth**).

The phallus of Heaven, the male force, reached out to the young girl, his daughter, Earth. Agni, the god of fire, had made the hot passion and seed of Heaven. As the act of union was committed, some seed spilled onto Earth, and words and the rituals were born. The god had satisfied himself with his own daughter. Out of the hot spilled seed came the Angirases, the mediators between gods and humans, who distribute the gifts of the gods. Heaven is our father and Earth is our mother.

Another Rig Veda myth, also from the tenth book, emphasizes the traditional creation myth theme, the sacrifice of a being out of whose dismemberment the world is made (see also **Babylonian Creation; Creation by Sacrifice; Icelandic Creation**).

The thousand-headed, thousand-footed primal man, Purusa, enveloped earth and was, in fact, the universe, the here and the there, the now and the always. Three quarters of Purusa are of the undifferentiated immortal sphere, one quarter is the sphere of life's forms.

When the gods performed the sacrifice of the primal man, his bottom quarter became the world we know. His mouth, out of which came words, became the wise brahmin priest and the god Indra. His arms became the warrior caste, his thighs the common people, and his feet the lowest of the low (see also **Animism**). Out of the sacrifice of Purusa came the beasts, plants, rituals, sacred words, and the Vedas themselves. From Purusa's mind came the moon, from his eye the sun, from his breath the wind, from his head the sky, from his feet the earth, from his navel the atmosphere.

By sacrificing Purusa, the gods sacrificed he who gave of himself. In so doing they made sacrifice to sacrifice and gave birth to the foundation of our order.

In the tenth book of the Rig Veda there is still another hymn of creation. It speaks of the necessity of opposites. Without Non-Being there cannot be Being; without Being there cannot be Non-Being. It is a human cry for knowledge of origins, for the meaning of Self, the meaning of Being. "Who knows the source of this creation?" the poet asks.

If in the beginning there was neither Being nor Non-Being, neither air nor sky, what was there? Who or what oversaw it? What was it when there was no darkness, light, life, or death? We can only say that there was the One, that which breathed of itself deep in the void, that which was heat and became desire and the germ of spirit.

The wise say that Non-Being and Being became one and that chaos became order. Who really knows what happened, whether it came from the One or not? Only the creator knows, and maybe even he knows nothing about it.

The following is a modern translation of the Rig Veda that attempts to capture the original style.

1 Then even nothingness was not, nor existence.
There was no air then, nor the heavens beyond it
What covered it? Where was it? In whose keeping?

Was there then cosmic water, in depths unfathomed?

[2] Then there were neither death nor immortality,
nor was there then the torch of night and day.
The One breathed windlessly and self-sustaining.
There was that One then, and there was no other.

[3] At first there was only darkness wrapped in darkness.
All this was only unillumined water.
That One which came to be, enclosed in nothing,
arose at last, born of the power of heat.

[4] In the beginning desire descended on it—
that was the primal seed, born of the mind.
The sages who have searched their hearts with wisdom
know that which is, is kin to that which is not.

[5] And they have stretched their cord across the void,
and know what was above, and what below.
Seminal powers made fertile mighty forces.
Below was strength, and over it was impulse.

[6] But, after all, who knows, and who can say
whence it all came, and how creation happened?
The gods themselves are later than creation,
so who knows truly whence it has arisen?

[7] Whence all creation had its origin,
he, whether he fashioned it or whether he did not,
he, who surveys it all from highest heaven,
he knows—or maybe even he does not know.

Translation by A. L. Basham. Reprinted from *The Wonder That Was India*, London, 1954, 247–248.

The Hindus continue their meditation on creation in later sacred texts called the Brahmanas, composed in the first millennium B.C.E. In a sense they are later commentary on the Vedas, especially the Rig Veda. As in the earlier creation hymns, the principle of heat is crucial, indicating a connection with other Indo-European or Aryan mythologies, especially that of Persia (Iran) (see also **Finnish Creation; Zoroastrian Creation**). The primal cosmic egg found here is also found in many Indo-European mythologies (see also **Greek Creation**).

One myth, from the Satapatha Brahmana, tells how in the beginning there was only the primeval sea—the waters. It was the waters who wished to reproduce, and through devotions became heated enough to produce a golden egg that floated about for a time. Then from the egg came Prajapati.

It took a year for him to come, and so it takes about that amount of time for a woman or a cow to give birth. After he broke out of the egg, Prajapati rested on its shell for another year or so before he tried to speak. The sound he made—the Word, his sounded breath—became earth. His next sound became sky. Other sounds became the seasons (see also **Creation by Word; Creation from a Cosmic Egg**).

After waiting another year Prajapati stood in his shell. He could see even then from the beginning of his life to its end in one thousand years.

Prajapati gave himself the power of reproduction. Some say he created the fire god, Agni, out of himself. With his hot breathing up into the sky *(div),* he created the gods *(devas)* above, and now there was light, *(diva).* With his breathing down Prajapati created the Asuras and the darkness of the earth. To avoid a cosmic struggle between light and dark, Prajapati overcame the Asuras with evil. Now there were, however, day and night.

Prajapati realized that by creating beings he had created time, and we know that Prajapati is, in a sense, time.

In the Aitareya Brahmana, Prajapati and his daughter (the sky or dawn) are characters in a reworking of the incest myth from the Rig Veda. We note the stronger presence of the taboo associated with the act.

Prajapati came to his daughter as a stag; she had the form of a doe. The gods watched and were horrified that Prajapati was doing "what is not done," and they created the monstrous and wild Rudra to punish him.

Rudra struck Prajapati with his arrow, and the progenitor became the constellation we call Capricorn or Deer's Head. In the piercing of Prajapati, his seed spilled and became a lake. The gods called this lake "not to be spoiled," so the seed of man is not to be spoiled. The gods gave the seed heat and Agni made it flow, and it became the Aditya (sun gods), the cattle, and many other things.

Still another Brahmana, the Kausitaki Brahmana, tells how, when Prajapati wanted to have offspring, he practiced deep asceticism that generated such heat as to give birth out of himself to fire, the sun, the moon, the wind, and the female dawn. Prajapati ordered the five to practice asceticism, and they did. Dawn took the form of a beautiful nymph, and when they saw her, the other four were so moved that their seed flowed. When they told their father, he made a golden bowl in which to collect the seed so it would not be lost. Out of the bowl of seed sprang a thousand-armed god, who took hold of his father and demanded a name before he would eat food. Prajapati named him Bhava (existence).

The next series of Hindu scriptures is called the Upanishads (learning sessions), in which thinkers of the period between about 800 and 400 B.C.E. gave thought to the earlier writings. The creation theories of the Upanishads quite clearly develop from those of the earlier texts and contain familiar themes.

In the oldest of the Upanishads, the Chandogya Upanishad, we still find the predominance of reproduction, heat, and primal waters in a reworking of the story of Prajapati and the golden egg. Prajapati has now become the creator god Brahma.

There was only Non-Being in the beginning. Non-Being developed into an egg. After a year the egg broke into two parts, one silver, one gold. The silver part is earth; the gold part is the sky. The various inside parts of the egg are the mountains, rivers, clouds, and so forth. The sun was born from the egg. At his birth, everything rose toward him. The sun is Brahma.

In the same Upanishad there is a version of the myth that postulates a world beginning not in Non-Being but in Being, for how could Being emerge from Non-Being? Being wanted to reproduce itself, so it gave off heat, which in turn procreated itself and gave off water. This is why we perspire today. The water in turn procreated itself and gave off food, which, of course, comes from water.

The Barhadaranyaka Upanishad returns to the early Vedic figure of Purusa, the primal man. It says that in the beginning was Atman (Soul—the One within) in the form of Purusa. When he looked about and saw nothing, he said "I am." He was lonely, however, so he became two, a husband and a wife. They came together and mankind was born. Then the wife became frightened and turned herself into a cow to hide; the man became a bull, found her, and cattle were born. So it went for the many forms of animals.

In a sense, Atman is Brahman within. In the Kena Upanishad the question of the nature of the ultimate One, Brahman, is considered. Brahman should not be confused with the creator god Brahma, although Brahma, like everything else, is in Brahman—specifically, his personified masculine form. Brahman itself is beyond gender or any other kind of definition or comprehension.

What spirit awakens the mind, makes life begin, and makes us speak in words? Who is the spirit behind seeing and hearing? "It is the ear of the ear, the eye of the eye, and the word of words, the mind of mind, and the life of life." Brahman cannot be spoken, but the spirit behind the possibility of speaking is Brahman.

The *Lawbook of Manu*, written in the second century B.C.E., is one of several books that continues the tradition of developing and commenting on scripture. The *Lawbook of Manu* takes its name from the tradition of Manu, the first human in each age of existence. Again, in this myth we find the motif of the primal egg and procreation. We also find the element from **creation by thought.**

The Self-Existent Brahman thought of the waters, and they were. His seed in the waters became a golden egg, and out of the egg Brahman was born as progenitor of all. Out of Brahman came the male Purusa, who is also Brahma. As for Brahman, it remained in the egg for a year before dividing the egg by thought into heaven and earth.

After 300 C.E., the process of reexamining scripture continued in the books called Puranas (old stories). Among other issues, these works consider the whole question of the identity of Brahman as revealed in the various gods. The most famous of the Puranas is the Vishnu Purana. It contains an **earth-diver creation** myth.

Brahma, the form taken by Brahman, the god without beginning or end, awoke and saw the empty universe. There were only the waters, the progeny of the eternal Brahman. Brahma decided that the earth lay beneath the waters. As at the beginning of each preceding creation he took the form of an animal—in this case a wild boar. In this form, based on the sacrifices of the sacred Vedas, Brahman as Brahma, the Great Boar, dove to the bottom of the primal waters to find Mother Earth. She received him with joy and hymns of praise, recognizing him as the creative principle behind all that was. "No one knows your true Being," she sang, "but everything that the mind can conceive or the senses perceive is a form of you." Brahman is Brahma, Vishnu (the great preserver god), and Shiva (the destroyer god). He is the soul of souls. Brahman can only be worshipped in the many forms it takes, since Brahman is formless.

The Great Boar came to the earth, and the wise praised him as the source of all things. He raised up the earth to where it floats now, a "mighty vessel" on the surface of the waters. Then Brahman in the form of Brahma created the world. He does this at the beginning of each *kalpa,* each creation that follows each world dissolution. He gives form to the power within the things to be created; that is, he gives form to Brahman.

There are many stories of Brahma as creator. In the epics—particularly the Mahabharata—he is depicted in more personal, less abstract terms than in the more self-consciously religious Puranas. Brahma, in the Mahabharata, is given human qualities such as jealousy. Called Prajapati again, or Grandfather, he creates wanton women for the purpose of stirring up men and bringing them to desire and anger. In this way men would be deluded and prevented from usurping the position of the gods. In another part of the Mahabharata, Brahma creates the woman, Death, to preserve the distinction between men and gods (see also Buddhist Creation; Dhammai Creation; Jain Creation; Minyong Creation).
Sources: O'Flaherty, 25–55; Leach, 221–223; Sproul, 187–191.

Indonesian Creation

See Ceram Creation; Yami Creation.

Inktomi

Inktomi (Spider), sometimes Iktome or Inktonmi, is a popular Indian **trickster** figure (see also **Assiniboine; Sioux Creations**). He is seen usually as a small, wiry man in tight leggings and is often the companion of another trickster, Coyote. Inktomi sometimes participates in creative acts, but he is more commonly associated with the erotic and bawdy side of the trickster, erotic urges being, after all, the source of human procreation. He is our precivilized, untamed natural urges. As such, he tricks ignorant girls into sexual acts, has a dream of his penis growing to an enormous length, and is himself tricked into sleeping with his own wife, whom he has neglected in favor of others. He is the frequent center of dirty jokes in the Sioux culture.
Source: Erdoes, 358, 372, 381.

Ipurina Creation

The Ipurina of Brazil have a creation myth dominated by a flood. The flood took place when the sun, which is, in fact, a tub of boiling water, tipped over.
Source: Freund, 10.

Iroquoian Creation

Iroquois is the name of a league of five nations—the Cayuga, Mohawk, Oneida, Onondaga, and Seneca nations—in the upstate New York and Ontario region of North America. The Iroquois are linguistically related to the Cherokee and share many mythical traditions with the Huron. Fragments of a rich Iroquois mythology remain with us, and there are specific creation myths belonging to the various Iroquois groups (see also **Onondaga Creation; Seneca Creation**). The myth given here was generally Iroquoian rather than peculiar to any one of the nations. It has earth-diver aspects and a creatrix who falls from the sky—familiar elements of Native American mythology (see also **Cherokee Creation; Earth-Diver Creation; Huron Creation**).

In the beginning there was only the primordial waters full of water birds and the Tortoise. The Ongwe lived in the heavens. The Ongwe father pushed his wife and their daughter out of a hole in the sky, and they fell for a long time through the emptiness between the heavens and the waters. The water birds saw her, and one of them succeeded in finding earth under the waters for her to land on. They placed it on Tortoise's back, and he became the foundation for the earth on which Star Woman landed and with which she began the process of creation.
Source: Weigle, 70.

Islamic Creation

See Muslim Creation.

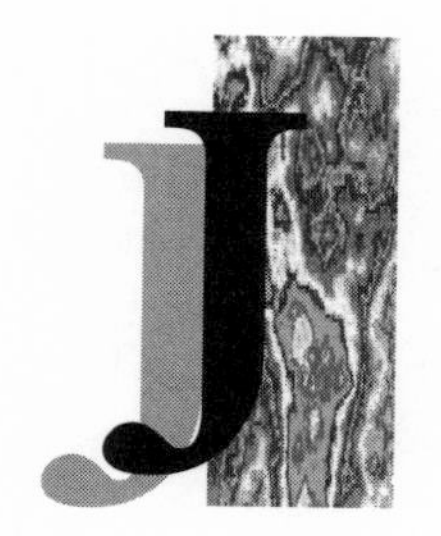

Jain Creation

Jainism is an ancient Indian religious system that took its present form in the sixth century B.C.E. It is based on a spiritual heroism of sorts in which the individual becomes a victor *(jina)* over worldly attachments. His soul is freed by strict ascetic practices rather than by prayer. Gods as such are of little importance to Jains. What might be called their non-creation creation myth is contained in the ninth-century Mahapurana by the Jain teacher Jinasena (see also **Indian Creation**).

Those who suppose that a creator made the world and mankind are misled. If God is the creator, where was he before creation, and how could a non-material being make anything so material as this world? How could God have made the world with no materials to start with? Those who suggest that God made the material first and then the world are trapped in a chicken-or-egg question. Those who say raw materials came about naturally are just as trapped, because we might just as well say that the world could, therefore, have created itself and have come about naturally.

If it is suggested that God created the world by a simple act of his will, one would have to ask how so perfect a being could have willed the creation of something else?

There are many arguments to support the idea that it is foolish to assert that the world was created by God. Why, for instance, would God kill his own creations, and if it is suggested that some of his creations are evil, one must ask why God would have created them in the first place, he being said to be perfect.

The fact is: the world is as uncreated as time; it has no beginning and no end; it exists through its own being and is divided into heaven, earth, and hell.

Source: Sproul, 192–194.

Japanese Creation

The primary sources for Japanese creation myths are the Kojiki (Records of Ancient Matters) collected in 712 C.E. by Futo no Yasumuro, and the Nihongi (Chronicles of Japan) compiled in 720 C.E. Both works are influenced by Chinese thought, and both reflect the animistic approach of the Shinto religion, the worship of the divine forces and forms in nature (see also **Animism**). The Chinese principles of yin and yang can be seen as personified in the Japanese figures Izanami and Izanagi, the first ancestors in a **creation from chaos** marked by the separation of world parents (see also **Creation from Division of Primordial Unity; World Parent Creation**).

According to the Kojiki, there was a time when there was only chaos until Heaven and Earth separated. At that time the Three High Deities created the passive and active principles, Izanami (Female who Invites) and Izanagi (Male who Invites). These two are the first ancestors, the makers, and the basis of all creation.

Upon entering the light, Izanagi washed his eyes, and the sun and moon were released. By bathing in the primeval sea, the gods of earth and sky were released.

The beginning as described in the Nihongi is more elaborate. In fact, the Nihongi creation is a tragic tale containing many themes found in other mythologies. There is a graphically phallic coital relationship between Heaven and Earth, for instance, and an Orpheus-like journey into the underworld to search for a lost loved one.

In ancient times, Heaven and Earth were still one. The In (yin) and Yo (yang) were still not separated. There was only an egglike chaos containing the seeds of creation. Heaven was made of the purer part of the mass; the heavy part was Earth. So Heaven raised himself first and the islands of earth began to form. Then between Heaven and Earth grew a strange plantlike form that became a great male god, followed by two others, also male. All were formed by the will of Heaven (see also **Separation of Heaven and Earth**).

Next, six deities were formed, and then Izanami and Izanagi.

The first ancestors in heaven wondered what lay below. So it was that they thrust down the jeweled spear of Heaven and stirred it about in the sea. As they lifted it, the liquid on the tip of the spear formed the island Onogoro-jima (Spontaneously Conceived Island). The deities descended to the island and built a land, with Onogoro-jima as its central pillar (see also **Yggdrasil**). Izanami and Izanagi wished to marry, so they devised a plan of courtship whereby they would walk in opposite directions around the world axle, the great pillar, until they met. Then Izanami said, "What a

Izanagi and Izanami, brother and sister of the Japanese creation myth, watch the first land form from ocean water dripping from Izanagi's spear.

beautiful youth I have met!" Izanagi objected, however, that it should have been he, the man, who spoke first, so they began the process again. When they met he said, "What a beautiful maiden I have met." Then Izanagi asked Izanami how her body was, and she said there was a place in her body that was empty and was the basis of her femininity. Izanagi said there was a part of his body that was superfluous but was the source of his masculinity. Perhaps if the superfluous and the empty, the masculine part and the feminine part, could join, procreation would be possible. Thus Izanami and Izanagi became one as husband and wife. Out of their union came many islands, and finally they created the Great Eight Island Country (Japan) and the sun goddess (Amaterasu or Ohohiro-me no muchi) as queen of the universe. Amaterasu was so radiant that her parents sent her to heaven to rule there. Then they produced the moon god to be their daughter's consort. Some of the children these two produced were dangerous—especially the Impetuous One and the god of fire. The first was exiled to the land of Yomi (the underworld) and the second burned his mother to death, but not before she gave birth to the water goddess, Midzuhano-me, and the earth goddess, Haniyama-hime, whom the fire god took as a wife.

Izanagi went to Yomi in search of the dead Izanami, but he arrived too late; she had already eaten food cooked in Yomi. She ordered her husband not to look at her, but Izanagi lit a torch and saw his mate in her state of putrefaction. In anger at his disobedience and at his having shamed her, Izanami and the Ugly Females (Furies) of Yomi chased Izanagi all the way to the land's entrance. Once he had escaped, Izanagi was plagued by bad luck, as he had visited the land of the dead. After being cleansed by sacred waters, Izanagi isolated himself forever on a distant island. Izanami became queen of the underworld.

> Of old, Heaven and Earth were not yet separated, and the In and Yo not yet divided. They formed a chaotic mass like an egg, which was of obscurely defined limits, and contained germs. The purer and clearer part was thinly diffused and formed Heaven, while the heavier and grosser element settled down and became Earth. The finer element easily became a united body, but the consolidation of the heavy and gross element was accomplished with difficulty. Heaven was therefore formed first, and Earth established subsequently. Thereafter divine beings were produced between them.
>
> Reprinted from a translation by W. G. Aston, London, 1924.

Sources: Eliade (B), 94; Olcott, 23; Sproul, 210–215.

Jewish Creation

See Hebrew Creation.

Jivaro Creation

The Jivaro people of Equador are a warlike and agricultural people best known for their practice of shrinking the heads of their war victims. Theirs is not a happy mythology; it is dominated by a series of battles among the gods. Their creation story is related to the **creation from chaos** type of myth.

At first there were the creator, Kumpara, and his wife, Chingaso. Their son was Etsa, the sun. One night Kumpara placed some mud in his mouth and spit it onto his son. In this way the girl Nantu, the moon, was conceived. This complicated manner of conception was accomplished so Etsa could marry Nantu, who was not born of his own mother.

Nantu resisted Etsa's advances, however, and escaped to the sky, where she painted herself in dark sombre colors and designs. A bird, Auhu, who also was enamored of Nantu, saw her leaving and tried to follow her, but Nantu cut down the vine on which he was trying to reach the sky.

Etsa was extremely angry when he discovered Nantu's escape, and he decided to follow her. Pulled up by a parrot on each wrist and a parakeet on each knee, he caught Nantu and fought violently with her, causing eclipses. When she had been overcome, Nantu wept. This is why it rains when the moon's face gets red.

Nantu now went off alone and produced her own child, Nuhi, by breathing on dirt. The jealous bird, Auhu, broke the clay child and the remains became the earth. Lonely now, Nantu finally agreed to be Etsa's wife. They had a son, Unushi the sloth, who was the first Jivaro. Unushi was given the forest on earth as a living place.

Etsa and Nantu went on reproducing, always using earth as a mating place. They produced the various animals.

The woman, Mika, was born of a mysterious egg sent by Chingaso. Mika was given to Unushi in marriage, and Etsa and Nantu instructed them on how to live. Unushi was lazy, however, and did little work. To this day the Jivaro women work harder than the men.

Unushi and Mika took a canoe down the river. A son, Ahimbi the water snake, was born to them in the canoe. Out of other eggs sent by Chingaso there came other animals and birds.

Some of the birds helped Mika and her husband find food, and the anaconda made them an axe. Ahimbi used it to cut down a tree and make his own canoe; he wanted to be on his own. He traveled about and eventually came back to see his parents. Mika was alone, however; Unushi had wandered off into the forest and got lost. That night Ahimbi slept with his mother and did not wake up soon enough to prevent Etsa from seeing them together (see also **Incest in Creation Myths**). In great anger Etsa exiled the couple, and they wandered about producing offspring and looking for shelter. The animals were disgusted by their incest, however, and would not help them. Unushi, upon learning what Mika and his son had done, blamed Nantu, beat her, and buried her in a hole. She escaped with Auhu's help, but since she was not grateful to him, he still cries in the night for her return.

Nantu told what had happened to her and the sons of Mika, and Ahimbi killed Unushi. Then Mika killed the sons for having killed her husband. The struggle between Mika and her sons caused thunder, lightning, and violent rain. Out of the loudest clap of thunder and a bolt of lightning sprang the fully armed Jivaro warrior, Masata (War), who from then on stirred up all the people and gods against each other.

Etsa and Nantu blamed Ahimbi for the trouble that had come to the earth. Etsa imprisoned him under the great falls, where, as indicated by the turbulent waters there, he is still trying to get free. Now Ahimbi wishes for peace; he sends up sprays of water to make rainbows so Etsa will see that he means well. Masata, however, always obscures the sign by sending mist or rain to block it. Once Chingaso came in a canoe wanting to rescue Ahimbi, but the snake, not recognizing the goddess, turned over her canoe in the rough water and ate her. So the wars go on.

Source: Sproul, 308–313.

Joshua Creation

The Joshua Indians of Oregon have an earth-diver myth that is a good example of **imperfect creation** (see also **Earth-Diver Creation**).

In the beginning, before there was land—only the sky, flat sea, and fog—two men lived in a sweat house on the water. One of these was the creator, Xowalaci; he stayed inside most of the time while his friend sat outside watching. One day the friend saw what looked like land approaching over the sea. It had two trees on it. Finally the land struck the sweat house, stopped, and began to expand. The watcher went in and told the creator, but the land was not yet solid, so the creator smoked a pipe. Then he blew smoke on the land, which stopped the land's motion and caused the flowers to bloom and grass to grow.

The creator made five mud cakes. One of these became a stone; he dropped it into the water and listened for it to get to the bottom. He kept dropping mud cakes until the land came near to the surface. Then the waves came; as they receded Xowalaci scattered tobacco seed and sand appeared. Then the breakers came, again leaving more land. Xowalaci stomped on it and it became hard land.

The creator saw a man's tracks and knew they meant trouble. He brought the water over the new land to rid it of the man, but the tracks always came back (see also **The Flood and Flood Hero**). Indeed, there has always been unexpected trouble in the world.

Now it was time to make people, so Xowalaci tried mixing grass and mud and rubbing it. He created a house and placed the two first figures in it. In a few days a dog and a bitch appeared, and later the bitch gave birth to puppies.

Xowalaci tried adding white sand to the mix to make people, but this time he made snakes. Out of some bad dogs he made water monsters. Still he could not make people.

The creator thought and thought until his friend suggested that he might have a try at creating. He suggested that he smoke the tobacco that night to see if anything would come out of the smoke. After three days of the friend's smoking, a house appeared and soon after that a beautiful woman. The woman was lonely, so Xowalaci gave his companion to the woman and proclaimed that their children would be the people.

To get the man and woman together, Xowalaci made the woman sleep. Then the man went to her. In a dream she experienced a man with her, but in the morning there was no man.

Soon a child was born. The woman longed for the father; she took the child on her back and searched for ten years for her husband. She neglected the child during the search and he almost died. They returned home and the boy asked where his father was. She explained to her son about the dream.

Now Xowalaci told his friend that the woman was home, and that night the woman wanted her husband badly. When the door opened she saw the man of her dreams. The first family was together. The creator explained things to them and did some final creating. The family had many children, who spoke different languages and formed the different tribes. Finally, Xowalaci left for the heavens, and the people stayed here.

Source: Sproul, 232–236.

Kagaba Creation

The Kagaba people of Colombia were created *ex nihilo* by a supreme deity who was female, a semi-abstract figure who represents the fecundity of earth. She is the Great Mother (see also **Creation from nothing; Mother-Creatrix**).

The Great Mother who bore us in the beginning is the mother of all that is. She made our songs, our seed, our nations. She is the mother of thunder, the mother of trees, the mother of all that grows and lives. She is the mother of the old stone-people and the mother of the young people, the French. She gave us the dances and the temples. When we sow the fields we say the Our Mother prayer: "Our mother of the growing fields, our mother of the streams, will have pity upon us. For whom do we belong? Whose seeds are we? To our mother alone do we belong."

Source: Eliade (B), 16.

Kakadu Creation

As in all Australian aboriginal myths we find the theme of the *ex nihilo* creative walkabout and the mysterious Dreaming of the particular world of the people in question. In this myth, however, it is the feminine power—the Great Mother, Imberombera—rather than the masculine power that acts out **the Dreaming** (see also **Creation from Nothing; Mother-Creatrix**).

The giant Wuraka walked through the western sea and came to Allukaladi between the Roe and Bidwell mountains, which he made. Then

he moved on to other places in our land. Meanwhile, Imberombera also arrived from the sea at Malay Bay or Wungaran. She met up with Wuraka and asked him where he was going. He said he was heading through the bush to the rising sun in the east. To speak to each other they used the language of the people of Port Essington. Wuraka carried his enormous penis over his shoulder, and it made him so tired that he sat down to rest rather than going along with Imberombera as she suggested. Where he rested, Tor Rock rose up. As for Imberombera, her belly was full of children, and she carried bags of yams and a large stick. At Marpur, near Wuraka's resting spot, she planted yams and left some spirit children who spoke Iwaidja. She went on to many places, leaving spirit children and different languages.
Source: Sproul, 323–325.

Kalevala

The *Kalevala* is the Finnish national epic, compiled in 1835 by Elias Lonnrott. It is based on ancient myths of the Finno-Ugric peoples (Finns, Lapps, and Magyars) and contains the **Finnish creation** myth.

Kato Creation

The Kato Indians of northern California say that Thunder and his companion, Nagaitcho, created the world from chaos (see also **Creation from Chaos**). They began by repairing the old sandstone sky and stretching it with rocks that formed the four directions, also making a clear path for the sun. They made clouds by making fires on hills and mist by making fires in valleys.

Then Thunder made a human. He took some earth and made the arms and legs, and he used some wadded up grass for the belly and heart. He used clay for the liver and kidneys and a reed for the trachea. For blood he used ochre and water. Finally, Thunder made the man's genitals and his eyes, nose, and mouth. Out of one of the man's legs he made a woman.

Now Thunder made it rain and he broke open rocks and trees with his power. There was a flood, and many people and animals died *(see also* **The Flood and Flood Hero**). The water animals—the whale, sea lion and others—were saved, however. Thunder placed redwoods along the new shores and mountains, as well as in other places where there was fresh water for

the people and animals to drink. He made many animals—good and bad—the bears, rattlesnakes, deer, and so forth. Thunder took his dog and wandered up and down the coast admiring his good, green, well-stocked earth. Then he went back north to his home, fully satisfied with his work.
Source: Thompson, 30.

Keres (Keresan) Creation

The Keresan people make up most of the southern pueblos along the Rio Grande in New Mexico. These groups—including the pueblos at Laguna and the Sky City (Acoma) in the east, and Santa Ana, Zia, San Felipe, Santo Domingo, and Cochiti, further west and north, near Santa Fe—are Keres-speaking and Christianized. Along with the new religion brought by the Spanish in the sixteenth century, however, the Pueblo people continue to maintain their original religious beliefs, myths, and ceremonies.

The Keresan creation stories are dominated by a female figure, **Thinking Woman,** or Sus'sistinako (Tsitctinako or Tsichtinako); although at the Zia pueblo she is male. Thinking Woman (also sometimes called Thought Woman or Prophesying Woman) in some ways resembles the sometimes Hopi generatrix, **Hard-Beings Woman** or Hurúing Wuhti (see also **Hopi Creation**). She also has characteristics of **Spider Woman,** the creatrix who is so important in Southwestern mythologies in general. The Keres creation is an emergence myth (see also **Creation by Emergence**) in which Thinking Woman, who is of the fertile womblike underworld (like Spider Woman), is able to carry her creative thought into the outside world. She has been called "a kind of silent Logos which brings everything into existence." The Acoma pueblo has a particularly lively version of the Thinking Woman creation (see also **Acoma Creation; Laguna Creation**).
Sources: Bierhorst, 82; Tyler, 82; Weigle, 32–33, 214–218.

Khepri

Khepri—there are several variations of the spelling—is an early version of the Egyptian creator god (see also **Atum; Creation from Secretion; Egyptian Creation**). He could be the soul, he who comes from the waters, the lotus, the hawk, the divine child, or any number of other forms, but he is always the high god.

Kiowa Creation

The Kiowa people live in southwestern Oklahoma. Their myth is a somewhat unique emergence myth (see also **Creation by Emergence**). It is said that the Kiowa emerged into this world long ago through a hollow log, far north of their present home, and that a pregnant woman got stuck in the log, blocking the way for others.

I

You know, everything had to begin, and this is how it was: the Kiowas came one by one into the world through a hollow log. They were many more than now, but not all of them got out. There was a woman whose body was swollen up with child, and she got stuck in the log. After that, no one could get through, and that is why the Kiowas are a small tribe in number. They looked all around and saw the world. It made them glad to see so many things. They called themselves *Kwuda,* "coming out."

II

They were going along, and some were hunting. An antelope was killed and quartered in the meadow. Well, one of the big chiefs came up and took the udders of that animal for himself, but another big chief wanted those udders also, and there was a great quarrel between them. Then, in anger, one of these chiefs gathered all of his followers together and went away. They are called *Azatanhop,* "the udder-angry travelers off." No one knows where they went or what happened to them.

III

Before there were horses the Kiowas had need of dogs. That was a long time ago, when dogs could talk. There was a man who lived alone; he had been thrown away, and he made his camp here and there on the high ground. Now it was dangerous to be alone, for there were enemies all around. The man spent his arrows hunting food. He had one arrow left, and he shot a bear; but the bear was only wounded and it ran away. The man wondered what to do. Then a dog came up to him and said that many enemies were coming; they were close by and all around. The man could think of no way to save himself. But the dog said: "You know, I have puppies. They are young and weak and they have nothing to eat. If you will take care of my puppies, I will show you how to get away." The dog led the man here and there, around and around, and they came to safety.

IV

They lived at first in the mountains. They did not yet know of Tai-me, but this is what they knew: There was a man and his wife. They had a beautiful child, a little girl whom they would not allow to go out of their sight. But one day a friend of the family came and asked if she might take the child outside to play. The mother guessed that would be all right, but she told the friend to leave the child in its cradle and to place the cradle in a tree. While the child was in the tree, a redbird came among the branches. It was not like any bird that you have seen; it was very beautiful, and it did not fly away. It kept still upon a limb, close to the child. After a while the child got out of its cradle and began to climb after the redbird. And at the same time the tree began to grow taller, and the child was borne up into the sky. She was then a woman, and she found herself in a strange place. Instead of a redbird, there was a young man standing before her. The man spoke to her and said: "I have been watching you for a long time, and I knew that I would find a way to bring you here. I have brought you here to be my wife." The woman looked all around; she saw that he was the sun.

V

After that the woman grew lonely. She thought about her people, and she wondered how they were getting on. One day she had a quarrel with the sun, and the sun went away. In her anger she dug up the root of a bush which the sun had warned her never to go near. A piece of earth fell from the root, and she could see her people far below. By that time she had given birth; she had a child—a boy by the sun. She made a rope out of sinew and took her child upon her back; she climbed down upon the rope, but when she came to the end, her people were still a long way off, and there she waited with her child on her back. It was evening;

the sun came home and found his woman gone. At once he thought of the bush and went to the place where it had grown. There he saw the woman and the child, hanging by the rope half way down to the earth. He was very angry, and he took up a ring, a gaming wheel, in his hand. He told the ring to follow the rope and strike the woman dead. Then he threw the ring and it did what he told it to do; it struck the woman and killed her, and then the sun's child was all alone.

VI

The sun's child was big enough to walk around on the earth, and he saw a camp nearby. He made his way to it and saw that a great spider—that which is called a grandmother—lived there. The spider spoke to the sun's child, and the child was afraid. The grandmother was full of resentment; she was jealous, you see, for the child had not yet been weaned from its mother's breasts. She wondered whether the child were a boy or a girl, and therefore she made two things, a pretty ball and a bow and arrows. These things she left alone with the child all the next day. When she returned, she saw that the ball was full of arrows, and she knew then that the child was a boy and that he would be hard to raise. Time and again the grandmother tried to capture the boy, but he always ran away. Then one day she made a snare out of rope. The boy was caught up in the snare, and he cried and cried, but the grandmother sang to him and at last he fell asleep.

> Go to sleep and do not cry.
> Your mother is dead, and still you feed
> upon her breasts.
> Oo-oo-la-la-la-la, oo-oo.

VII

The years went by, and the boy still had the ring which killed his mother. The grandmother spider told him never to throw the ring into the sky, but one day he threw it up, and it fell squarely on top of his head and cut him in two. He looked around, and there was another boy, just like himself, his twin. The two of them laughed and laughed, and then they went to the grandmother spider. She nearly cried aloud when she saw them, for it had been hard enough to raise the one. Even so, she cared for them well and made them fine clothes to wear.

VIII

Now each of the twins had a ring, and the grandmother spider told them never to throw the rings into the sky. But one day they threw them up into the high wind. The rings rolled over a hill, and the twins ran after them. They ran beyond the top of the hill and fell down into the mouth of a cave. There lived a giant and his wife. The giant had killed a lot of people in the past by building fires and filling the cave with smoke, so that the people could not breathe. Then the twins remembered something that the grandmother spider had told them: "If ever you get caught in the cave, say to yourselves the word *thain-mom,* 'above my eyes.'" When the giant began to set fires around, the twins repeated the word *thain-mom* over and over to themselves, and the smoke remained above their eyes. When the giant had made three great clouds of smoke, his wife saw that the twins sat without coughing or crying, and she became frightened. "Let them go," she said, "or something bad will happen to us." The twins took up their rings and returned to the grandmother spider. She was glad to see them.

IX

The next thing that happened to the twins was this: They killed a great snake which they found in their tipi. When they told the grandmother spider what they had done, she cried and cried. They had killed their grandfather, she said. And after that the grandmother spider died. The twins wrapped her in a hide and covered her with leaves by the water. The twins lived on for a long time, and they were greatly honored among the Kiowas.

X

Long ago there were bad times. The Kiowas were hungry and there was no food. There was a man who heard his children cry from hunger, and he went out to look for food. He walked four days and became very weak. On the fourth day he came to a great canyon. Suddenly there was thunder and lightning. A voice spoke to him and said, "Why are you following me? What do you want?" The man was afraid. The thing standing before him had the feet of a deer, and its body was covered with feathers. The man answered that the Kiowas were hungry. "Take me with you," the voice said, "and I will give you whatever you want." From that day Tai-me has belonged to the Kiowas.

XI

A long time ago there were two brothers. It was winter, and the buffalo had wandered far away. Food was very scarce. The two brothers were hungry, and they wondered what to do. One of them got up in the early morning and went out, and he found a lot of fresh meat there on the ground in front of the tipi. He was very happy, and he called his brother outside. "Look," he said. "Something very good has happened, and we have plenty of food." But his brother was afraid and said: "This is too strange a thing. I believe that we had better not eat that meat." But the first brother scolded him and said that he was foolish. Then he went ahead and ate of the meat all by himself. In a little while something awful happened to him; he began to change. When it was all over, he was no longer a man; he was some kind of water beast with little short legs and a long, heavy tail. Then he spoke to his brother and said: "You were right, and you must not eat of that meat. Now I must go and live in the water, but we are brothers, and you ought to come and see me now and then." After that the man went down to the water's edge, sometimes, and called his brother out. He told him how things were with the Kiowas.

Kodiak Creation

The Eskimo **creation from chaos** myth of Kodiak Island in Alaska is one of the many Eskimo Raven creations (see also **Eskimo Creation;**

Raven). It contains the excretory aspects so common to **trickster** stories (see also **Creation by Secretion**).

Raven brought light from the sky, and at the same time a bladder, containing a man and a woman, came down. By pushing and stretching it, the man and woman made the bladder into the world. By pushing with their hands and feet, they made mountains. Trees came into being when the man scattered his hair about. The woman urinated and spit to make the oceans, lakes, rivers, and ponds. The man made a knife out of one of the woman's teeth and cut some wood to make woodchips, which he then threw into the water to make fish. The man and the woman had a son, who played with a stone that became an island. They left another son on that island with a female dog that became his wife. These were the ancestors of the Kodiak people.

Source: Bierhorst, 61.

Kono Creation

In Guinea, the Konos say in their **creation from chaos** myth that Sa (Death) came before anything. It is rare to have Death as the primary creator.

Once there was only darkness, and Sa lived there with his wife and daughter. Sa wanted something more stable, so he used his magic to make a slushy kind of mud sea as a place to live. The god Alatangana suddenly appeared—where from, we do not know—and visited Sa. Alatangana was disgusted by Sa's life-style and told him so. He decided to do better than Sa had done.

Alatangana first made the slush solid and added animals and plants to it to make it more lively. Sa was pleased, and he offered friendship to the god and entertained him. When Alatangana asked Sa for his daughter's hand in marriage, however, he refused.

Alatangana met the girl secretly, and they eloped to a distant place to avoid the anger of Sa. There in the faraway place they produced 14 children: four white boys, four white girls, three black boys, and three black girls. The children all spoke different languages, and their parents could not understand them. This was so upsetting to Alatangana and his wife that the god decided he would have to consult his father-in-law, Sa.

Alatangana went to Sa and asked his advice. Sa was not pleased with his son-in-law, who had eloped with his only daughter, and he admitted that it had been he who had punished the disobedient couple by making it impossible for them to understand their own children. "Still," he said, "I will give your white children intelligence and the ability to write, so that they can write down what they want to say. I will give your black children tools so that they may feed and shelter themselves. But the black children must marry only black people and the white children only whites."

Alatangana agreed to his father-in-law's rules, and so began the black and white races and their separation. The people scattered over the world. The world was still covered in darkness, though, so Alatangana had to once again get help from Sa, who told two birds how to sing in a way that would bring light.

These birds, the tou-tou and the cock, became morning birds, whose song does indeed bring light. The light came from the sun, whose course Sa had fixed. He also arranged the moon and stars for night light.

Then Sa called Alatangana to him and demanded a service in return for the gifts he had given. "You have stolen my only child," he said. "Now you must give me one of yours whenever I wish it. When I wish to call one of your children I must never be denied. You will know I have called by the sound of the calabash rattle in your dream." So it was that death for us humans was the bride-price for Alatangana's marriage.

Source: Beier, 3–6.

Kootenay Creation

The Kootenay (Kutena) Indians of the North American Plains believed that Coyote was the creator, that he created the sun, for instance, out of a ball of grease. This seems a tricksterlike way to create (see also **Coyote**).

Source: Olcott, 5.

Krachi Creation

The Krachi people of Togo in Africa tell a creation myth that includes the Fall (see also **Fall of Humankind**) with elements of the separation of the primordial unity between earth and heaven (see also **Creation from Division of Primordial Unity**).

In the beginning Wulbari (heaven—male) lived on top of Asase Ya (earth—female). Man lived between them, but with little room to move. Man's squirming irritated Wulbari so much that he left and went up above.

One of the things that bothered Wulbari was an old woman who, when grinding maize, kept hitting him with her pestle, and the smoke from the cooking fires bothered his eyes. Some say that Wulbari was annoyed because

men would sometimes wipe their dirty hands on him. The Krachi people say that there was an old woman who used to cut off bits of Wulbari to flavor her soup.
Source: Sproul, 75.

Kukulik Creation

The Eskimos of Kukulik Island in the Bering Strait say that Creator-Raven first made the Unisak Cape shoreline, then the Russian part of the world, then the American. He rested for a bit and then, in earth-diver fashion, reached down into the deep water for some sand (see also **Earth-Diver Creation; Raven**). He squeezed the water out of the sand and made the village of Cibukak (Wrung Out) from it. Out of the pebbles in the sand he made the people, and he told them how to pick seaweed and hunt sea animals and fish.

One day a man decided to travel to the Sun to ask him for reindeer for the people. Sun refused, but he did give the man certain pebbles, which he said should be thrown into the water. The man did as he was told, and when the stones hit the water they became whales. The Kukulik Eskimos are still whalers.
Source: Leach, 196.

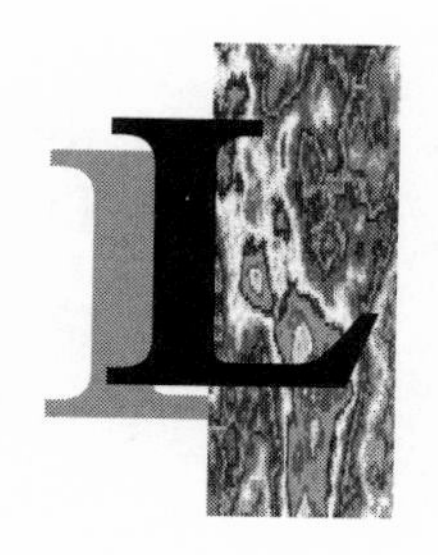

Laguna Creation

The Laguna pueblo near Acoma, New Mexico, is the home of one of the Keresan peoples. As in the cases of their relatives, the Laguna tell an emergence/*ex nihilo* creation myth in which the major part is played by **Thinking Woman,** Old Spider Woman, or Tse che nako, as they call her (see also **Acoma Creation; Creation from Nothing; Keresan Creation**).

In the beginning Thinking Woman made everything, even thoughts and the names of things. Then her children, the girl twins Uretsete and Naotsete, made more names and thoughts.

> Ts'its'tsi'nako, Thought-Woman,
> is sitting in her room
> and whatever she thinks about
> appears.
>
> She thought of her sisters,
> Nau'ts'ity'i and I'tcts'ity'i,
> and together they created the Universe
> this world
>
> and the four worlds below.
> Thought-Woman, the spider,
> named things and
> as she named them
> they appeared.

She is sitting in her room
thinking of a story now
I'm telling you the story
she is thinking.

Source: Weigle, 19.

Lakota Sioux Creation

See Sioux Creation.

Lapp Creation

See Finnish Creation.

Leach, Maria

Maria Leach is a scholar of creation myths who compiled *The Beginning: Creation Myths around the World* (see also **Thompson, Stith**).

Lenape (Delaware) Creation

The Lenape *ex nihilo* creation story is found in the Walam Olum, or Red Book (see also **Creation from Nothing**). It tells the story of a perfect creation undermined by a great magician serpent and an ensuing flood. On the opposite page is a pictorial version of the Red Book.

Lisa (Liza)

A popular African creator god (see also **Fon Creation**).

1. Sayewitalli wemiguma wokgetaki.
Sayewi — At first
talli — there
wemi — all
guma — Sea water
Wokget on the top
aki — Land.

1st glyph

2. Hacking-kwelik owanaku wakyutali
Kitanitowit-essop.
Hacking — above
Kwelik — much water
Owanaku — foggy (was
wak — and
yutali — there
Kitanitowit — God Creator
Essop — he was

3. Sayewis hallemiwis nolemiwi elemamik
Kitanitowit essop
Saye-wis — first being
hallemi-wis — eternal being
nolemiwi — invisible
elemamik — every where
Kitanitowit essop — God Creator he was

144969

3

The Lenape (Delaware) of what is now the eastern United States used pictographs on sticks to record their beginnings. In 1820 a copy of the glyphs with the Lenape interpretation came to naturalist and writer Constantine Rafinesque, who translated them into English. The first glyph represented land covered by water.

Long, Charles H.

A student of the myth and religion scholar Mircea Eliade (see also **Eliade, Mircea**), Charles H. Long is the author of *Alpha: The Myths of Creation*. In this work he categorizes creation myths according to basic types (**Creation from Chaos, Creation from Nothing,** etc.).

Lugbara Creation

The Lugbaras are a large tribe in Uganda and Zaire. The Lugbara live in small clan groups and place a high premium on lineage.

The people are all of one blood. The creator, Adronga 'ba o'bapiri, made this blood *ex nihilo* for the first two creatures he placed here (see also **Creation from Nothing**). These were Gborogboro (sky person—male) and Meme (big body-female). Meme was full of wild animals, who sprang from her womb.

Gborogboro and Meme, a form of heaven and earth, were married. After Meme gave birth to the animals, the creator placed children in her womb or, as some say, the pair coupled and children were conceived—a boy and a girl. This boy and girl made another male and female pair, and that pair did the same.

It is said by some that the various brother-sister pairs never copulated but that the women became pregnant when goat's blood was poured onto their legs. People also say that eventually humans and the creator were separated and that the people, although carrying the same blood, were separated into black and white.

Source: Middleton, 47–51.

Luiseno Creation

See Shoshonean Creation.

Magyar (Hungarian) Creation

See Finnish Creation.

Mahabharata

Probably the world's longest literary work, the Mahabharata is the great epic of India and a primary source for Hindu mythology. Versions of the Mahabharata existed long before it began to be transcribed in the early fifth century B.C.E., and it continues to be expanded to this day. Like an ancient tree and like Hinduism itself, it sheds a branch here and slowly grows a new one there. It continues to send its seeds on the wind even into the less poetic media of modern India and South Asia. It has taken root in movies, puppet theaters, and even comic books of modern New Delhi and Bombay. With its myth of creation, tales of heroism, rules for social behavior, and deeply philosophical scripture (the Bhagavad Gita), the Mahabharata is in every sense both a bible and a living epic, the record of a creation in process (see also **Indian Creation**).

Maidu Creation

Baptismal-like cleansings and re-creations mark this **earth-diver creation** myth of the California Maidu Indians. The myth is of interest because it provides a basis for Maidu rituals, centered in the strange figure of Father of the Secret Society, who is clearly shamanic. We also find the familiar figure of **Coyote** in this myth (see also **Trickster**).

In the beginning there was no light, and everywhere there was water. From the north a raft came carrying Turtle and Pehe-ipe (Father of the Secret Society). A feather rope was let down from the sky. Earth Starter came down the rope, tied it to the bow of the raft, and climbed aboard. His face was masked and he shone as if he were the sun itself; he sat still and was quiet.

"Where did you come from?" Turtle asked.

"From up there," said Earth Starter.

"It would be nice if you could make me some dry land to stand on once in a while," said Turtle. "And, by the way, are there going to be people on the earth?"

"Yes," answered Earth Starter.

"When?"

"I can't say. But if you want land I will need some earth."

Turtle said he would dive for some. Earth Starter tied a rock around Turtle's arm. He then took a feather rope out of the blue and tied it around Turtle's leg.

"Good," said Turtle. "If the rope is not long enough I'll jerk once. If it is long enough I'll jerk twice and you must pull me up." Then he dove over the side as Father of the Secret Society shouted.

Turtle did not come up for six years, and when he did he was covered in slime and had only a bit of earth under his nails. Earth Starter took a stone knife from under his armpit and scraped the bit of earth from Turtle's nails. He rolled it in his hands and placed it on the stern of the raft. Gradually the little ball grew until it was as big as the world and had grounded the raft at Ta'doiko-o.

"This is fine," said Turtle, "but we need light; can you make some?"

"Well," said Earth Starter, "let's see. Come out onto the raft and I will call my sister from the east."

The light began to rise, and Father of the Secret Society shouted again. Earth Starter made a path for his sister, Sun, and after a while she went down at the other end. Father of the Secret Society was upset, so Earth Starter called his brother, Moon, and he came up and all was well.

"Will you do nothing else?" asked Turtle.

"Oh, yes," answered Earth Starter, and he called out the stars by name and made the huge Hu'kimtsa (oak) tree to grow at Ta'doiko-o. They all sat under it for two days (see also **Yggdrasil**). Then they went off to look at Earth Starter's new world. Earth Starter went so fast, however, that the others could only see a ball of fire flashing under the ground and in the water.

Meanwhile, back at Ta'doiko-o, Coyote and his dog, Rattlesnake, came up out of the ground. Some people say that only Coyote could see through Earth Starter's mask to his face. All five beings now built huts for themselves, but it was forbidden to go into Earth Starter's hut.

For some time Earth Starter made other things—birds, trees, and deer. Sometimes Turtle complained about Earth Starter's style and methods.

One day Earth Starter and Coyote were at Marysville Buttes when suddenly Earth Starter announced that he would make people. He took red clay from the earth and made a man and a woman. He laid them next to each other inside his house, where no one else had been allowed to go. Then Earth Starter lay down next to the two new beings. He stretched out his arms and sweated for at least a day and night until early in the morning, when the new woman tickled him. He did not laugh, but got up and struck the ground with pitch wood so a fire came. The new people were as white as snow and had pink eyes and black hair. Earth Starter finished by giving the people hands like his so they could climb trees to escape bears. These first people were called Ku'ksu and Morning Star Woman.

Now Coyote thought he would have a try at creating people, since Earth Starter had explained to him how it was done. In the morning, however, when the woman he created tickled him, he laughed—something Earth Starter had told him not to do. As a result, these new people had glassy eyes and did not come alive. "I told you not to laugh," said Earth Starter, and Coyote said he had not. This was his first great lie.

Soon there were many people. Earth Starter did not stay around as much as he had before, but he did speak to the first man, Ku'ksu, during the night sometimes. One night he ordered him to gather the people the next day and take them to the little lake nearby. He said that Ku'ksu would be an old man by the time he got there.

Ku'ksu did as he was told, and when the people got to the lake he was an old man. He fell into the lake and sank into it. There was a terrible roaring sound and earth-shaking until Ku'ksu came out of the water as a young man. Then Earth Starter spoke to the people and told them that when they got old they must do as Ku'ksu had done and all would be well. After he had spoken, Earth Starter returned to his place above.

The world that Earth Starter had left was a perfect place. There was more than enough to eat. In fact, the women put out baskets at night and found them full of warm food in the morning.

Coyote introduced new ways, however, and everything changed. Some of the new ways were all right; there were games and races. The bad thing that Coyote introduced was death.

It was up to Ku'ksu to teach the people how to treat the dead by wrapping them and burying them "until the world shall be made over." He then sent the people to different places, where they spoke different languages, and he went to the spirit house. Coyote tried to follow him there by killing himself, but by then Ku'ksu had gone above to where Earth Starter is.
Source: Thompson, 24–30.

Malozi Creation

In Zambia, the Malozi believe that it was Nyambe who was first in the world. They say he lived on earth with his wife, Nasilele, and that it was he who made the rivers, plains, animals, and the first people, Kamunu and Kamunu's wife *ex nihilo* (see also **Creation from Nothing**). Like so many creation stories, however, this is the story of how humans fell (see also **Fall of Humankind**) and how they became separated from God and then built something like the Tower of Babel to get back to him.

Kamunu learned quickly from Nyambe and was more intelligent than the other animals. He learned to carve and to forge iron. A time came when he went so far as to forge a spear and kill the son of the antelope, which he ate.

Nyambe was furious. "You have eaten one of my children," he said. "This was your own brother." He then sent Kamunu away. After a year, Kamunu came back with a club and a magic pot. This was reported to Nyambe, who allowed the man to stay.

Kamunu went to Nyambe, demanded fields to cultivate, and was given them. At night, however, the buffalo trampled the fields, so Kamunu killed one. Nyambe said he could eat it and he did.

Then Kamunu's magic pot died, and Nyambe said that was the way of the world.

When Kamunu killed a deer who trampled his fields he was allowed to eat it. The next day his dog died, but Nyambe was not concerned for him.

Kamunu now claimed that he had seen Nyambe with both his pot and his dog, but his wife did not believe him.

When Kamunu killed a trampling elephant, Nyambe said he could eat it. When he did, however, his child died. When Kamunu complained, Nyambe said "This is the way of the world."

Then Nyambe took his messenger and the antelope and went away from Kamunu to live on an island. Kamunu found Nyambe, built a canoe, and took animal offerings to him to eat. Nyambe was unhappy and would not receive the dead offerings. "These are all my children," he said.

Nyambe went to a mountain to escape the man, but Kamunu followed him there. Wherever Nyambe went on the earth he found Kamunu's children. Finally Nyambe found a safe place. He invited the animals to come with him to get away from the man and his children. They decided to stay, however, thinking they could use their speed, strength, and size to protect themselves. In the end, Nyambe went up on high. Still Kamunu tried to get to Nyambe. He built a tower of wooden posts tied with bark to try to get to Nyambe, but the tower fell from the weight of the climbers and many people died. Kamunu has given up trying to get to Nyambe, but every morning he prays to him when the sun, our king, comes up.
Source: Beier, 7–14.

Mandan Creation

The Mandan Indians, who probably originated in the eastern area of what became the United States, settled in North Dakota. They were ravished by diseases brought by whites, and eventually all but died out. Their otherwise typical **earth-diver creation** myth is clearly influenced by the religion of Catholic missionaries who converted them in the eighteenth century. The figure of Lone Man is related to Christ, as is the world tree, which is at the center of Mandan ceremonies (see also **Yggdrasil**).

First there was water and darkness everywhere. There was the Creator and his companion, Lone Man. As they were walking about on the waters, the Creator and Lone Man met a small being, which turned out to be a duck. They asked her how she managed to live, and she explained that she took sustenance from something under the waters. Creator asked her to show him the food, and she dove to the depths and returned with sand.

The Creator and Lone Man decided to create land from the sand. Leaving some water between them—the Missouri River—they began creating. North of the river the Creator made lakes; northern animals like the elk, deer, and antelope; deep valleys; and high mountains. There was ample material for food and shelter for both humans and animals. Farther south, Lone Man made plains-like country with a few lakes and rivers, along with animals such as the beaver, muskrat, and cattle of different kinds.

Then the two discussed their creations. The Creator did not much like Lone Man's flat and somewhat forbidding landscape; it was not as suitable as his for providing food, clothing, and shelter. The Creator's animals seemed better adapted to the elements. Lone Man suggested that the people could make use of the Creator's superior creation first but that later his might become useful—and so it happened.

The rest of the story tells how Lone Man entered the world of humans as a savior and teacher.

This is what an old Mandan woman in the 1920s had to say about Lone Man's time with the people.

> In the course of time Lone Man looked upon the creation and saw mankind multiplying and was pleased, but he also saw evil spirits that harmed mankind and he wanted to live among the men that he had created and be as one of them. He looked about among all nations and peoples to find a virgin to be his mother and discovered a very humble family consisting of a father, mother and daughter. This virgin he chose to be his mother. So one morning when the young woman was roasting corn and eating it he thought this would be the proper time to enter into the young woman. So he changed himself into corn and the young woman ate it and conceived the seed. In the course of time the parents noticed that she was with child and they questioned her, saying, "How is it, daughter, that you are with child when you have not known man? Have you concealed anything from us?" She answered, "As you say, I have known no man. All I know is that at the time when I ate roast corn I thought that I had conceived something, then I did not think of the matter again until I knew I was with child." So the parents knew that this must be a marvel since the child was not conceived through any man, and they questioned her no more.
>
> In course of time the child was born and he grew up like other children, but he showed unusual traits of purity and as he grew to manhood he despised all evil and never even married. Everything he did was to promote goodness. If a quarrel arose among the people he would pacify them with kind words. He loved the children and they followed him around wherever he went. Every morning he purified himself with incense, which fact goes to show that he was pure.
>
> The people of the place where he was born were at that time Mandans. They were in the habit of going to an island in the ocean off the mouth of a river to gather *ma-ta-ba-ho.*

For the journey they used a boat by the name of *I-di-he* (which means Self Going); all they had to do was to strike it on one side and tell it to go and it went. This boat carried twelve persons and no more; if more went in the boat it brought ill luck. On the way to the island they were accustomed to meet dangerous obstacles.

One day there was a party setting out for the island to get some *ma-ta-ba-ho* and everyone came to the shore to see them off and wish them good luck. The twelve men got into the boat and were about to strike the boat on the side for the start when Lone Man stepped into the boat, saying that he wanted to go too. The men in the boat as well as the people on the shore objected that he would bring ill luck, but he persisted in accompanying them and finally, seeing that they could not get rid of him, they proceeded on the journey.

Now on the way down the river, evil spirits that lived in the water came out to do them harm, but every time they came to the surface Lone Man would rebuke them and tell them to go back and never show themselves again. As they neared the mouth of the river, at one place the willows along the bank changed into young men who were really evil spirits and challenged the men in the boat to come ashore and wrestle with them. Lone Man accepted the challenge. Everyone with whom he wrestled he threw and killed until the wrestlers, seeing that they were beaten, took to their heels. Then he rebuked the willows, saying that he had made them all and they should not turn themselves into evil spirits any more. When they reached the ocean they were confronted by a great whirlpool, into which the men in the boat began to cast trinkets as a sacrifice in order to pacify it, but every time they threw in a trinket Lone Man would pick it up saying that he wanted it for himself. Meanwhile, in spite of all they could do, the whirlpool sucked them in closer. Then the men began to murmur against Lone Man and complain that he brought them ill luck and lament that they were to be sucked in by the whirlpool. Then Lone Man rebuked the whirlpool saying, "Do you not know that I am he who created you? Now I command you to be still." And immediately the waters became smooth. So they kept on the journey until they came to a part of the ocean where the waves were rough. Here the men again began to offer sacrifices to pacify the waves, but in spite of their prayers and offerings the waves grew even more violent. And this time Lone Man was picking up the offerings and the men were trying to persuade him not to do so, but he kept right on,—never stopped! By this time the boat was rocking pretty badly with the waves and the men began to murmur again and say that Lone Man was causing their death. Then he rebuked the waves, saying, "Peace, be still," and all at once the sea was still and calm and continued so for the rest of the trip.

Upon the island there were inhabitants under a chief named Ma-na-ge (perhaps water of some kind). On their arrival, the chief told the inhabitants of his village to prepare a big feast for the visitors at which he would order the visitors to eat all the food set before them and thus kill them. Lone Man foresaw that this would happen and on his way he plucked a bulrush and inserted it by way of his throat through his system. So when the feast was prepared and all were seated in a row with the food placed before them, he told the men each to eat a little from the dish as it was passed from one man to the next until it reached Lone Man, when he would empty the whole contents of the dish into the bulrush, by which means it passed to the fourth strata of the earth. When all the food was gone, Lone Man looked about as if for more and said, "Well! I always heard that these people were very generous in feeding visitors. If this is all you have to offer I should hardly consider it a feast." All the people looked at the thirteen men and when they saw no signs of sickness they regarded them as mysterious.

Next Ma-na-ge asked the visitors if they wanted to smoke. Lone Man said "Certainly! for we have heard what good tobacco you have." This pleased Ma-na-ge, for he thought he would surely kill the men by the effects of the tobacco. So he called for his pipe, which was as big as a pot. He filled the pipe and lighted it and handed it over to the men. Each took a few puffs until it came to Lone Man, who, instead of puffing out the smoke, drew it all down the bulrush to the fourth strata of the earth. So in no time the whole contents of the pipe was smoked. Then he said he had always heard that Ma-na-ge was accustomed to kill his visitors by smoking with them but if this was the pipe he used it was not even large enough to satisfy him. From that time on Ma-na-ge watched him pretty closely.

(You may put in about the women if you want to).

Now Lone Man was in disguise. The chief then asked his visitors for their bags to fill with the ma-ta-ba-ho, as much as each man had strength to carry, and each produced his bag. Lone Man's was a small bag made of two buffalo hides sewed together, but they had to keep putting in to fill it. The chief watched them pretty closely by this time and thought, "If he gets away with that load he must be Lone Man!" So when the bag was filled, Lone Man took the bag by the left hand, slung it over his right shoulder and began to walk away. Then Ma-na-ge said, "Lone Man, do you think that we don't know you?" Said Lone Man as he walked away, "Perhaps you think that I am Lone Man!" Ma-na-ge said, "We shall come over to visit you on the fourth night after you reach home." By this he meant, in the fourth year.

When they reached home, Lone Man instructed his people how to perform ceremonies as to himself and appointed the men who were to perform them. He told them to clear a round space in the center of the village and to build a round barricade about it to take four young cottonwood trees as a hoop. In the center of the barricade they were to set up a cedar and paint it with red earth and burn incense and offer sacrifices to the cedar. Lone Man said, "This cedar is my body which I leave with you as protection from all harm, and this barricade will be a protection from the destruction of the water. For as Ma-na-ge said, they are coming to visit you. This shall be the sign of their coming. There will be a heavy fog for four days and four nights, then you may know that they are coming to destroy you. But it is nothing but water. When it comes, it will rise no higher than the first hoop next to the ground and when it can get no higher it will subside."

After he had instructed them in all the rites and ceremonies they should perform he said, "Now I am going to leave you—I am going to the south—to other peoples—and shall come back again. But always remember that I leave with you my body." And he departed to the south. And after four years Ma-na-ge made his visit in the form of water and tried in every way to destroy the inhabitants of the village, but when he failed to rise higher than the first hoop he subsided.

Reprinted from Martha Warren Beckwith, *Mandan-Hitatsa Myths and Ceremonies. Memoirs of the American Folklore Society,* New York: J. J. Augustin, for the American Folklore Society, [1937] 1938.

Source: Sproul, 248–250.

Mande Creation

The creation myth of the Mande people of Mali is an example of a world egg myth (see also **Creation from a Cosmic Egg**), and it has elements of the **imperfect creation** type. There is also the theme of incest (see also **Incest in Creation Myths**) and a sacrificial Christlike figure and a flood (see also **The Flood and Flood Hero**).

In the beginning Mangala (God) made the *balaza* seed. This seed did not work well, however, so he made two kinds of eleusine seeds and other pairs of seeds that would become the four directions, the four elements, and all of the things that organize creation. There were also seeds for two pairs of twins, each set made up of a male and female; these would become the first people. All of the seeds were together in God's egg, the world egg.

One of the male twins, Pemba, thought he could take dominion over creation, so he left the egg before his time. He fell through space until a part of his placenta became the earth. God made the other half the sun. Pemba was not pleased with the barren earth, however, and he tried to return to the egg. This was not possible, but Pemba did steal some male seeds from God. These he planted in the earth, an act of incest, because earth was made from his placenta and, therefore, from his own mother. One of the seeds, an eleusine, took root in the placenta's blood and came out red. The incest made the earth impure.

Meanwhile, back in heaven, the male twin called Faro, who took the form of a fish, was sacrificed to atone for Pemba's sin. His body was cut into sixty pieces, which became trees on earth, symbols of resurrection (see also **Animism; Creation by Sacrifice**). Then Mangala brought Faro back to life, gave him human form, and sent him to earth on an ark made of the primordial sacred placenta, the mother of all.

The ark landed on Mount Kouroula in the land of Mande, which means son of the fish. This was the World Mountain (see also the **Primal**

Mound). Faro stood tall on the ark; he had the original eight ancestors with him and all the first animals and plants. Like Faro, all the first beings contained the male and female life force in balance. The first beings watched the first sunrise from the ark.

Then the ancestor of bards, Sourakata, descended from heaven with the first sacred drum, the skull of the sacrificed Faro. He played on the drum to bring rain. No rain fell, so the primordial smith came down and struck a rock with his hammer, bringing rain. Now the twin fish (symbolizing Faro and Faro's "son") came down, too. This is why the mannogo fish is sacred to the Mande.

In the time that followed, Faro created the world as we know it out of Mangala's original egg seeds and their descendants. The first people continued the planting and built sanctuaries. Faro struggled against the evil force represented by Pemba, his evil brother. Faro, who is also the Niger River (his head is Lake Debo), is fertility itself, containing the male and female life forces. Faro flooded all of the land containing the impure seed of his brother. Only the good were saved in Faro's ark, and he taught them proper ways and how to make and keep the shrines.

Source: Sproul, 66–75.

Manu

Manu was the first man in some Hindu creation myths (see also **Indian Creation**).

Maori Creation

The Maori are a Polynesian people who have inhabited New Zealand since long before the arrival of Europeans. They possess a highly sophisticated and complicated religious and mythological system that concerns itself with profound spiritual matters and the nature of Being itself (see also **Hawaiian Creation; Polynesian Creation; Samoan Creation; Society Island Creation; Tahitian Creation**).

Io, or Iho, is the Supreme Being. He is the timeless *ex nihilo* creator of the universe, and the Maori creation myth is a metaphor for all types of creation, whether human or cosmic (see also **Creation from Nothing**). It is for this reason that the ritual words by which Io made the world are used to this day to help in the conception of a child, in the composition of a poem, or in the renewal of a broken mind or spirit.

The Polynesian people of New Zealand, the Maori, believed that Maui, shown pulling in a fish in this carved panel, was the son of Tangaroa, the creator. A hero to the Maori, Maui gave them fire.

In many parts of New Zealand the creation myth does not say much, if anything, about Io. Sometimes, like so many, it begins with a primordial unity that must be separated—differentiated—in order that creation can take place (see also **Creation from Division of Primordial Unity**). The Maori cosmogony usually begins, then, with the union of Rangi (Heaven) and Papa (Earth). There are several versions of this myth. The one that follows does speak of Io.

In the beginning there was darkness and water, where Io lived alone and inactive. In order to become active, Io uttered words calling on darkness to become "light-possessing darkness." So came light. When Io called for the light to become "dark-possessing light," darkness returned. Day and night had been born. Io continued creating with words—the "ancient and original sayings, which caused growth from the void" (see also **Creation by Word**).

Io called on the waters to separate and the heavens to be formed. Then Io became the gods. Most important, he created Rangi and Papa—Sky Father and the Earth Mother—who cleaved together in a procreative embrace, crowding their offspring. Two of these, Rongo and Tane, created plants, forests, and insects. Tane separated his parents to make more room; he was the god of life. Rangi and Papa were so sad to be separated that to this day Rangi drops tears on Papa and Papa's sighs rise as mist to her spouse.

Other children of the first parents were the winds, rains, earthquakes, and Tu, the warrior god whose children are the fearless Maori. The tenth child of Rangi and Papa was Tangaroa, the father of the hero Maui. Some say that the sun is the eye of Maui and that the eyes of his children became the evening and morning stars. Others say that Maui was thrown into the sea by his moonmother, Taranga, and rescued by Io, who hung him on the roof of his house.

The Maoris call New Zealand the Fish of Maui. Maui gave fire to humans, and he died in a search for immortality. He needed to make that search because the last child of Rangi and Papa brought death to the world.

The first Maori was made by the god Tane out of red clay (see also **Creation from Clay**). Some say that it was the god Tiki who made the first man in his own image, and thus he named him Tiki after himself.

The following is an example of a Maori creation chant:

> From the conception the increase,
> From the increase the thought,
> From the thought the remembrance,
> From the remembrance the consciousness,
> From the consciousness the desire.

The world became fruitful;
It dwelt with the feeble glimmering;
It brought forth night:
The great night, the long night,
The lowest night, the loftiest night.
The thick night, to be felt,
The night to be touched,
The night not to be seen,
The night of death.

From the nothing the begetting,
From the nothing the increase,
From the nothing the abundance,
The power of increasing
The living breath:
It dwelt with the empty space,
And produced the atmosphere which is above us,
The atmosphere which floats above the earth;
The great firmament above us dwelt with the early dawn,
And the moon sprung forth;
The atmosphere above us dwelt with the heat,
And thence proceeded the sun;
They were thrown up above,
As the chief eyes of Heaven:
Then the Heavens became light,
The early dawn, the early day,
The mid-day.
The blaze of day from the sky.

Reprinted from Richard Taylor, *Te Ika a Maui,* London: Wertheim and Macintosh, 1855, quoted in A. W. Reed, *Treasury of Maori Folklore,* Wellington, New Zealand: A. H. and A. W. Reed, 1963.

Sources: Eliade (C), 30; Leach, 172–174; Olcott, 29–30; Sproul, 337–346.

Marduk

Marduk was the dominant god of the Babylonians (see also **Babylonian Creation; Enuma Elish**).

Mariana Islands Creation

The Mariana Islanders of Micronesia tell of a creator called Na Arean, who creates *ex nihilo* by thought (see also **Creation by Thought; Creation from Nothing**). His assistant is his son, who performs another traditional chore of creators, the separation of heaven and earth, the primordial unity (see also **Creation from Division of Primordial Unity**).

In the beginning Na Arean was alone, "a cloud that floats in nothingness." He did not exist in our way because there was not yet any existence. Then a thought came from himself into his mind, and he made water in one hand and then mud, which he rolled and sat on as if it were an egg.

After a while his head swelled until, on the third day, a man broke out. Na Arean greeted this new great thought as it came alive. "You are to be called Na Arean the Younger and are to sit as you will in my right or left eye." This the little man did until Na Arean told him to go down to the earth he had made and find its center. "But where is the center?" the man asked. Na Arean the Elder took a tooth from his mouth and thrust it into the earth and said, "This is the center, the navel." Then Na Arean the Younger went down through the hollow tooth.

When he got to the center of the earth he found that things were all pressed together and dark, perhaps because his father had sat on the place. So he organized things until the heavens were where they belonged and the sun could send light through Na Arean's hollow tooth.

Source: Sproul, 336.

Marshall Islands Creation

This Micronesian *ex nihilo* creation story includes a report of the origins of Bikini Island, where the first atomic bomb test took place and the bikini bathing suit got its name.

In ancient times when there was only water, Lowa, the uncreated, was alone. When he made a humming sound the islands emerged, along with the reefs and sandbanks. He hummed again and the plants and animals arrived. Lowa made four gods for the four directions and a gull god to constantly circle the sky (see also **Creation from Nothing**).

Lowa also created a man, who put the islands into a basket made from coconut leaves and tried to set them in order—the Carolines to the west and the Marshalls in a straight line, except for Namorik, which fell out of the basket. After he had placed all the islands, he threw away the basket *(kilok),* and it became Kili Island.

Only one of the islands had coconut trees at the time. This was the little sandy island of Bikini (from *bok,* meaning sand, and *ni,* meaning coconut).

Lowa sent his tattooers down to the islands to give each creature its own mark.

Source: Leach, 185.

Mayan Creation

The Quiché Mayan people are the best known of the ancient Mayan peoples of the Guatemalan highlands. Their civilization dates back at least to 300 C.E., and their sacred history is preserved in the **Popol Vuh** or Book of the People. The Quiché creation is an example of an **imperfect creation** and of an *ex nihilo* creation by thought and word (see also **Creation by Thought; Creation by Word; Creation from Nothing**).

In the beginning there were only the creators, Tepeu and the Feathered Serpent, Gucumatz, in the void and the waters. These two sat together and thought. They glittered with sun power. Whatever they thought and whatever they said came into being. They thought the emptiness of the void should become something and it did. "Let there be earth," they said, and there was earth. They thought, "Mountains," and there were mountains. They said, "Trees," and there were trees. So it went. They separated the sky and earth (see also **Separation of Heaven and Earth**), and they made animals. They called on the animals to praise them, but the animals could not.

The creators thought they had better make beings who could be more aware of creation and of them. They wished to be praised. The first animals of creation they relegated to a low position. They would live outside in the wild and be hunted.

The creators spoke again to create beings they hoped would be able to praise them. They made a human form out of clay. It fell apart when it got wet, so they tried again with wood. These wood creatures did not fall apart. They walked and talked and made more of themselves, but they were too inflexible, too mindless, and without inner being. They did not think of their makers, and they caused troubles on earth, so Tepeu caused a great flood to destroy these imperfect beings. Those who were left were chased by the other animals into the woods to be monkeys.

The makers had to hurry; the dawn of our world was approaching. With the help of Mountain Lion, Coyote, Parrot, and Crow, they got together the things that would be food for the people—mainly corn, beans, and water—and made the four first people of the four directions.

These first four, the Quiché ancestors born of Tepeu and the Feathered Serpent, spoke well, worked well, and knew what there was to know. They

The Maya of Central America included the goddess Ixchel in the Popol Vuh, their sacred history. A clay ocarina from about 750 represents the goddess Ixchel in her aspect as Xquic with her twins Hunahpu and Xbalanque.

also praised their makers with the right words. These first four were wise and powerful. Tepeu and the Feathered Serpent feared them, so they removed some of the people's power and vision. Then they gave wives to the first four, and these original people procreated. They made the Quiché and the other tribes, too.

What follows is a segment from the Popol Vuh as translated by Dennis Tedlock.

> Now it still ripples, now it still murmurs, ripples, it still sighs, still hums, and it is empty under the sky.
>
> Here follow the first words, the first eloquence:
>
> There is not yet one person, one animal, bird, fish, crab, tree, rock, hollow, canyon, meadow, forest. Only the sky alone is there; the face of the earth is not clear. Only the sea alone is pooled under all the sky. . . .
>
> Whatever there is that might be is simply not there: only the pooled water, only the calm sea, only it alone is pooled.
>
> Whatever might be is simply not there: only murmurs, ripples, in the dark, in the night. Only the Maker, Modeler alone, Sovereign Plumed Serpent, the Bearers, Begetters are in the water, a glittering light. They are there, they are enclosed in quetzal feathers, in blue-green. . . .
>
> And then came his word, he came here to the Sovereign Plumed Serpent, here in the blackness, in the early dawn. He spoke with the Sovereign Plumed Serpent, and they talked, then they thought. . . .
>
> And then the earth arose because of them, it was simply their word that brought it forth. For the forming of the earth they said 'Earth.' It arose suddenly, just like a cloud, like a mist, now forming, now unfolding.

A modern Mayan creation myth was told to Morriss Siegel in the highlands of Guatemala in 1941. It is a Christian Mayan story.

> In a green pasture God was born of the Virgin Mother, and the ox came and looked and breathed upon him to give him life.

"Who is the father of our sister's child?" said the Ancient Men, the brothers of the Virgin Mother. "Whose son is this?"

Three days after God was born, he spoke. When the Ancient Men heard that, they said, "Come along," and they took him with them to put him to work. They gave young God an ax and told him to clear away the trees. God gave two chops with the ax and the whole field was cleared.

This was more than the brothers could do in a week, and they were afraid. They did not like this; they struck him with whips and they gave him no food when the day's work was done. They took him home that night to his mother and said, "Tomorrow we will take our brother to work again."

"No," said the Virgin Mother, "he is too small."

"But I want to go," said God. "I want to see them burn the field." It was the custom of the Maya Indians first to clear a field of trees and then to burn it off before planting.

So the next day God went along with the brothers. But when they came to burn the field, they tied God with a rope to a tree in the middle of it before they set fire to the brush, and then they stood in a circle around the fire, waiting. "His stomach will boil and burst," they said. They could not see him for the flames and the smoke.

When God saw the fire leaping toward him, he called to the little rodents in the earth, and the little rodents came out of their holes and gnawed the ropes and God escaped with them deep into their burrows, safe from the burning field.

"He is burned; we are rid of him," said the Ancient Men when the smoke lifted and they saw that he was gone.

But that night on their way home they saw God playing in front of his mother's hut. "We sent him home ahead," the Ancient Men told the Virgin Mother. "We were afraid he would get burned."

That night in their houses, the Ancient Men were unhappy and afraid.

"We have not conquered our sister's son," they said. "We must dance."

So the brothers brought out the drum and the chirimía (oboe). They put on their masks and costumes and then they danced.

"I am going to watch my brothers dance," said the Mother of God.

"I will go with you," said God.

"No, they will trample you."

"I can go in the cotton basket and watch."

So God went along in his mother's cotton basket and watched the dance of the Ancient Men.

When the dance was over, the brothers sat down to a feast.

"I want to see what they are eating," said God. So he looked. And when the men saw God watching, they threw the bones from their meat in his face.

God picked all the bones up very carefully and put them in his hat. He took them home and planted them in the earth nearby, and then built a corral around the place. In three days he went to look. Inside the corral he saw cattle and horses, pigs, deer, sheep, goats, armadillos, chickens, ducks, turkeys and geese, rabbits, squirrels, foxes, coyotes, badgers, raccoons, snakes, mice, birds, owls and songbirds and woodpeckers and sparrows, lizards, toads, frogs, iguanas, scorpions, turtles, cockroaches, dogs and cats, butterflies, ants, fleas and lice, and lions. Thus God planted all the animals there are in the world.

The Ancient Men came along and saw the corral and all the animals, so they opened it and most of the animals ran out. The horses, the sheep, the cattle, the pigs and chickens, the dog and the goat stayed behind in the corral. The wild animals ran into the forests and the birds flew away.

God said to his mother, "Now I am going to make a fiesta, a dance, like that of the Ancient Men."

But she laughed at him. "Not likely!" she said. "It takes money and food."

But God did it. "It will begin at midnight," he said.

Midnight came. The Ancient Men heard the skyrockets and ran to see the dance of God. It was better than theirs and they felt ashamed.

They said to young God, "We have been mean to you. Please forgive us and show us where you got those wonderful costumes." They wanted to throw away their old dance costumes and get new ones like God's.

"All right," said God, "but the place is far away. The traveling is hard. There is cramp on the road and bitter cold and need for prayer. But come, I will show you."

They went. The mountains were steep; the rocks were sharp; it was cold. Their legs cramped from the difficult climbing, and the Ancient Men stopped oftener and oftener to pray that they might endure it.

When they came to a big tree by the path, God said, "Climb it, my brothers," so the brothers went up.

The tree started to grow. The brothers went on climbing and the tree went on growing. And God willed that they should stay there and went home and left them.

"Where are the brothers?" said his mother.

"Far away," said God.

"Where's that?"

"I will show you," said God. When they came to the tree, it was full of monkeys.

"Is this you?" said the Virgin Mother. But they could not answer now. They leaped from branch to branch and looked at God and his mother. One threw down something and hit the Mother of God in the eye.

"Stay there!" she said. "It serves you right." So God and his mother went home. And this is why the monkeys are called "brothers of God."

But this was not the end of God's troubles. The foreign kings came running after him, and so God ran away.

On the way he saw a man planting. "What are you doing?" asked God.

"I am planting beans."

"In three days they will be ready," said God. "I am being hunted," said God. "Do not tell them I was here today." And he hurried off.

When the kings came to that place they saw a man gathering his beans.

"Have you seen God?" they said.

"Yes," said the man. "He passed by the day I was planting beans." The kings figured that was sixty days ago. But they followed God's trail.

After a while the kings caught him and killed him. Then they nailed him to the cross and went away.

God put a ladder on the cross and climbed up it into heaven. As soon as God arrived in heaven, a great light shone all over the world. There was the sun for the first time.

The cock crowed right away. The beasts in the forest howled; the cattle lowed. The wicked kings were burnt up by the sun's heat. And the world God had made was lighted with a great clear light.

Sources: Hamilton, 87–99; Leach, 98–104; Sproul, 288–298.

Melanesian Creation

See Banks Islands Creation; Fiji Islands Creation; New Guinea Creation; New Hebrides Creation; San Cristobal Creation.

Mesopotamian Creation

See Assyrian Creation; Babylonian Creation; Sumerian Creation.

Micronesian Creation

See Gilbert Island Creation; Mariana Island Creation; Marshall Island Creation; Truk Island Creation.

Miji Creation

See Dhammai Creation.

Minyong Creation

The Minyong are a non-Hindu people of northern India. Their creation myth contains the familiar theme of the **separation of heaven and earth** or world parents (see also **Indian Creation; World Parent Cre-**

ation). There are also similarities between this creation and the Amaterasu (sun goddess) aspect of the **Japanese creation.**

In the beginning there was the woman, Sedi, who was earth, and the man, Melo, who was sky. When they decided to get married, the creatures between them were afraid they might be crushed by the lovemaking of the great couple. A being called Sedi-Diyor hit Melo hard, and he retreated upwards from his earth-wife. Before he left, Melo gave Sedi two daughters. Sedi, however, was so sad at the departure of her husband that she refused to care for her children. She found a nurse for them, and they grew more radiant each day. When their nurse died and had to be buried, however, the girls died of grief and there was darkness everywhere.

The people were afraid of the darkness and missed the girls, so they dug up the nurse to see what might be the matter. In the grave they found only two great shining eyes in which they saw their own reflections. They washed the eyes but there were still images in them. They called a carpenter, who made models of the reflections. These turned out to be two girls.

The two girls grew in the house where they were kept. The older one became so bright that she left the house and began her life journey. The world became light as she entered it. When her bright sister followed her there was too much light, and things began to wither. The people decided they must kill one of the girls to lessen the heat and the light.

Frog agreed to do the deed; he shot the second girl, Bong, with an arrow, and the light and heat diminished. Rat took Bong's body to her sister, Bomong, who in her grief hid her head under a stone. The world grew dark again, and the people were afraid.

They sent Cock to find Bomong. He urged her to come out from under the rock, but she refused. She said she would come out only if her sister was revived, so the carpenter came again. He was able to put some of Bong's light back. She was, of course, the moon. Then Bomong, who is the sun, came back too, and everything was fine.
Source: Hamilton, 117–121.

Miwok Creation

See Pohonichi Miwok Creation.

Mixtec Creation

The Mixtec Indians are an ancient tribe living in the Mexican states of Oaxaca, Guerrero, and Puebla. Their **creation from chaos** myth tells

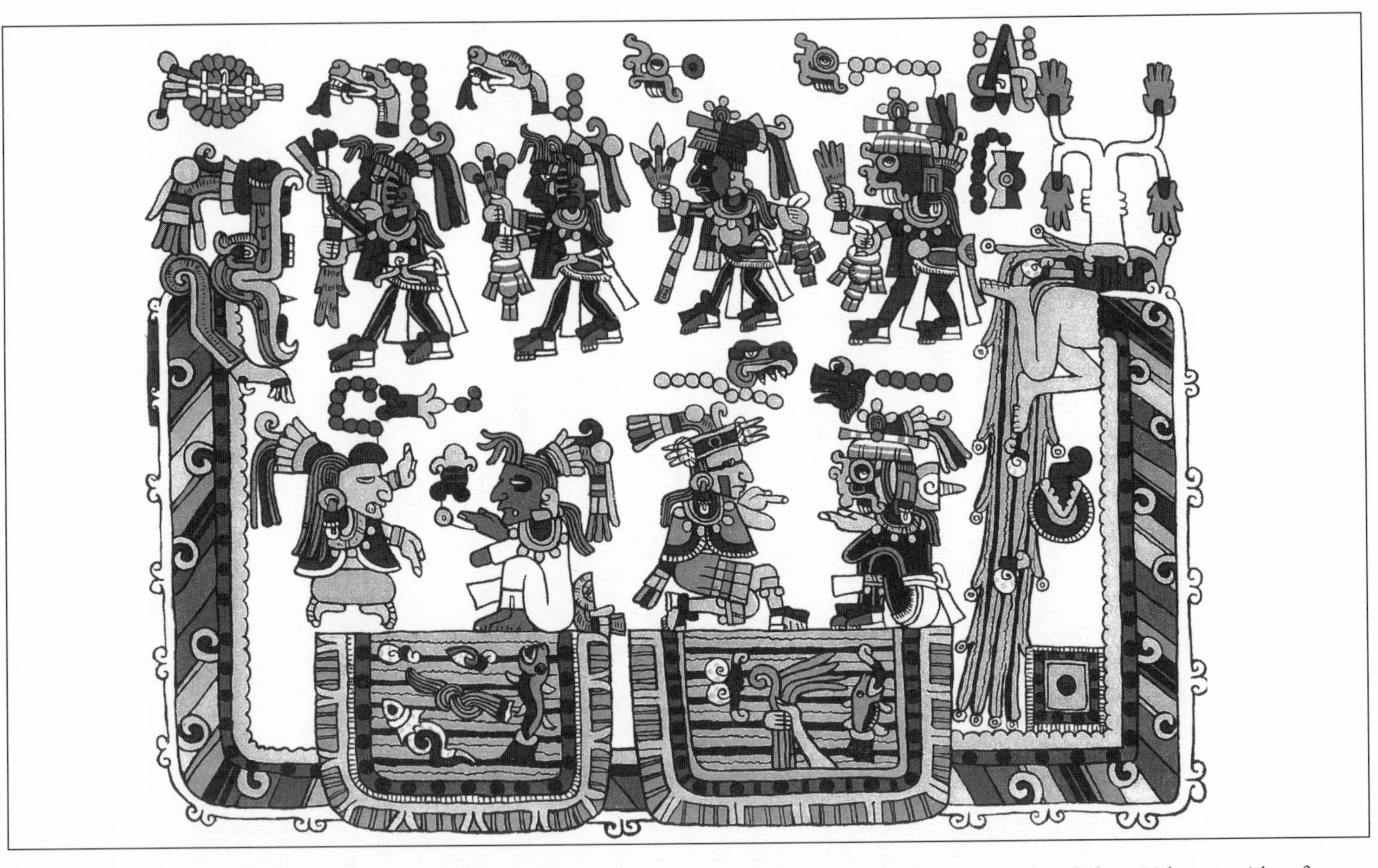

The Mixtec, a tribe of ancient Mexico, recorded their history on painted deerhide. This copy of the thirty-sixth page of the book shows the god Tloque Nahuaaque, right, a figure half man and half water. Other figures on the page, center, are also associated with water, an important element of Mixtec creation.

of a beginning when there was not yet time, when the earth was covered in water and darkness, and green slime was on the waters. Time began when the Mixtec god and his beautiful wife took human form and tamed the chaos, making a wondrous home for themselves on top of a cliff they raised over the waters. They rested the sky on a great ax they had placed on the cliff.

These first gods had two sons who played on the great cliff and had marvelous powers. They learned the arts of husbandry, and they used tobacco to pray to the gods for an earth on which people could plant things. The creator god agreed to their wish and freed the earth of the primeval sea so people could plant and roam about.

The cliff home of the gods is still in Oaxaca near the Mixtec river of origins where the first Mixtec were born from the trees.

Source: Leach, 93–94.

Modoc Creation

The Modoc Indians of Oregon were hunters and gatherers. There are two versions of the Modoc creation, one an emergence myth (see also **Creation by Emergence**) dominated by a god figure, one an **earth-diver creation** myth featuring a culture hero-trickster-shaman figure who may be the source or basis of a bear totem (see also **Trickster**).

Once Kumush, also called Old Man, went with his daughter to the spirit world beneath the earth. Down there in that beautiful place the spirits gathered each night to sing and dance. By morning they had returned to the spirit house and were bones again.

After a few days, Kumush decided to go back to the upper world; he wanted to take some spirits along to become new people. He tried to carry some bones away in a basket, but each time he tried, he tripped and fell, dropping the bones. Finally Kumush and his daughter managed to get to the upper world. They threw the bones down on the ground and cried out, "Indian bones," and the various tribes were formed. They threw bones toward the west and named them Shastas (brave warriors) and bones just a bit north and called them Klamaths (as easy to frighten as women). They continued in this manner until all the tribes were created. The bones nearby Kumush made into the Modoc people, a small but brave tribe.

Finally, Kumush told the people how to eat and how to divide work between men and women. Then Kumush and his daughter went to the place where the sun rises and traveled the sun road to the middle of the sky. They still live there today.

Others say that it was Kumokums who made the world, rather the way a child makes a world by playing in the sand. In the beginning there was only Kumokums and Tule Lake. He reached way down to the bottom of the lake and got some earth to make land around the lake. He took the earth and patted it, and it grew until he was sitting on a little island in the lake and the lake was surrounded by land. Then he played with the land, piling up mountains and scratching rivers out with his fingernails. He went on to pull trees out of the ground, and then he made birds and other animals. Then he was tired, so he slept through the winter under the lake. He made a little hole so he could look out once in a while to see how things were going. That hole is still there, and some day Kumokums will surely wake up.
Sources: Erdoes, 109–111; Marriot (A), 44–46.

Mongolian Creation

The central figure in the **creation from chaos** story of the Mongols is the Lama (superior one) of Tibetan Buddhists, who brought their point of view to Mongolia.

When there was no earth, but only water, the Lama descended from the heavens and stirred the waters. All of this stirring caused a congealing of the waters into land.
Source: Sproul, 218.

Monsters in Creation

Monsters in creation are common (see also **Babylonian Creation; Greek Creation**). They would seem to represent the early bestial aspects of creation before the "civilizing" influence of the given culture, represented by the high gods.

Mosetene Creation

For the Mosetene Indians of the Bolivian Andes, the world was created *ex nihilo* by Dobitt, whose home is heaven (see also **Creation from Nothing**). His world is a great raft that floats in space. It is guided by angel-like spirits. Dobitt populated the earth with humans, whom he made first out of clay (see also **Creation from Clay**).

Dobitt sent his son Keri as a white condor to see what was going on in the world, but the rope by which he was let down broke and Keri was killed. Dobitt made a fish out of his son's head, and then came back to the world himself to complete the creation. He made all the animals, poured out water from his basket to make rivers, lakes, and oceans, and taught humans how to survive.

Some say that Dobitt arranged for the sky to be held up above the earth (see also **Creation from Division of Primordial Unity**) by a huge serpent, but who knows, really?
Source: Leach, 127–128.

Mother-Creatrix

Civilizations whose religions grew primarily out of agricultural practices often conceived of the creator as female rather than male (see also **Divine Woman as Creator; Earth Mother; World Parent Creation**). Thus the earth is seen as the procreative, fruitful mother, the source of all bounty.

The Mound

See Primal Mound.

Munduruc Creation

The Munduruc Indians of Brazil believe that the creator is Karusakaibo, who was assisted in the beginning by Daiiru, the armadillo. Their myth is related to the emergence type of creation (see also **Creation by Emergence**).

One day Karusakaibo was so angry at Daiiru that the little armadillo hid in the ground. The creator blew into the hole and stomped his foot so hard that the culprit was blasted out. He told Karusakaibo that there were humans below the earth. The two friends, now reconciled, decided to get the humans up to the surface. They made a beautiful rope of cotton, such as the Mundurucs make today, and dropped it down the hole with Daiiru tied to one end. He showed the people how to climb out, but so many hung

onto the rope that it broke. Half of the people had to stay in the underworld, where they still are. Everything is the opposite there from what it is here; the sun goes in the opposite direction, it is night there when it is day here, and the moon is there when it is not here. The people came up at Necodemos, and the Mundurucs are the ones who look most like Karusakaibo.
Source: Leach, 123–124.

Muslim (Islamic) Creation

In the Koran, the holy book of Islam, dictated by the angel Gabriel to the prophet Mohammed in the seventh century C.E., we find several references to an *ex nihilo* creation (see also **Creation from Nothing**). At one point the book asks the people how they could not believe in a god (Allah) who was so powerful as to create the world in two days. In Sura (Section) XLI, the Koran says that Allah created heaven and earth by calling them to him "in obedience" (see also **Creation by Word**). We are told by the Koran that Allah created the world "to set forth his truth," that man was created from "a moist germ," and that the merciful God also created cattle and the fruits of the earth (Sura XVI). Elsewhere the book says that God created man "of the dust, then of the germs of life, then of thick blood" (Sura XL) (see also **Swahili Creation**).

> It is God who hath given you the earth as a sure foundation, and over it built up the Heaven, and formed you, and made your forms beautiful, and feedeth you with good things. This is God your Lord. Blessed then be God the Lord of the Worlds!
>
> He is the Living One. No God is there but He. Call then upon Him and offer Him pure worship. Praise be the God the Lord of the Worlds!
>
> Say: Verily I am forbidden to worship what ye call on beside God, after that the clear tokens have come to me from my Lord, and I am bidden to surrender myself to the Lord of the Worlds.
>
> He it is who created you of the dust, then of the germs of life, then of thick blood, then brought you forth infants: then he letteth you reach your full strength, and then become old men (but some of you die first), and reach the ordained term. And this that ye may haply understand.

The Muslim creation story is recorded in the pages of the Koran, which is rarely translated from Arabic. Allah's word came to the prophet Mohammed through the angel Gabriel.

It is He who giveth life and death; and when He decreeth a thing, He only saith of it, "Be," and it is.

[Sura XL, The Believer: 66–70]

Your Lord is God, who in six days created the heavens and the Earth, and then mounted the throne: He throweth the veil of night over the day: it pursueth it swiftly: and he created the sun and the moon and the stars, subjected to laws by His behest: Is not all creation and its empire His? Blessed be God the Lord of the Worlds!

Call upon your Lord with lowliness and in secret, for He loveth not transgressors.

And commit not disorders on the earth after it hath been well-ordered; and call on Him with fear and longing desire: Verily the mercy of God is nigh unto the righteous.

And He it is who sendeth forth the winds as the heralds of his compassion, until they bring up the laden clouds, which we drive along to some dead land and send down water thereon, by which we cause an upgrowth of all kinds of fruit.—Thus we will bring forth the dead. Haply ye will reflect.

In a rich soil, its plants spring forth a undantly by the will of its Lord, and in that which is bad, they spring forth but scantily. Thus do we diversify our signs for those who are thankful.

[Sura VII, Al Araf: 52–57]

Reprinted from a translation by J. M. Rodwell, London: J. M. Dent, 1909.

Muysca Creation

The ancient Muyscas of Peru and Colombia are said to have been created by the sun. They believed that the sun was Bochia, a very old bearded man who taught them about planting and rituals. Bochia's wife was jealous of the time he spent with the people and sent a flood to destroy them (see also **The Flood and Flood Hero**). The Sun set things to right, however; he made his wife the moon and he dried up the flood.

Source: Olcott, 26–27.

Nandi Creation

When the Nandi people arrived in what is now Kenya, they found the Dorobo people already there and soon assimilated their mythology. In fact, in the Nandi creation the Dorobo are the first people, who, as in so many African creations, are a source of tragedy for the rest of creation. It is of interest to note in this myth that God found a creation already there when he decided to set up the world as we know it, and that he seems to have had little to do with the way things eventually worked out.

God decided that it was time to arrange the world. He came down to earth and found the thunder, the elephant, and a Dorobo man already there. The thunder and the elephant were irritated by the fact that the Dorobo could turn over, seemingly at will, in his sleep while they had to get up to turn over.

In fact, the thunder was so upset by the man's behavior and so afraid of him that he fled to the sky, where he still lives. The elephant chided the thunder for fearing a being so small. "I fear him because he is evil," said the thunder. As for the man, he, as it happened, was afraid of thunder and was not displeased to see him retreat to the sky.

Then the man took a poison arrow and decided to kill the elephant with it. The elephant was terrified and begged the thunder to take him up to the heavens, but the thunder refused. "I thought you were not afraid of a being so small," he rumbled. The man killed the elephant and continues to do so. The man reproduced and took power all over the world.

Source: Sproul, 48–49.

Navajo Creation

The largest of the Native American nations is the Dine (People) or Navajos, as the Spanish first called them. They were, relatively speaking, latecomers to the Dinehtah, a vast homeland in the Four Corners region of the American Southwest. Recent scholarship suggests they may have arrived as early as 1000 C.E. They remained until they were expelled from the area in 1864 by a force led by Kit Carson and were forced to make the decimating Long Walk to Fort Sumner on the New Mexico-Texas border. There they lived the life of miserably treated prisoners for four years, and it was not until 1868 that those who were left were marched back to a greatly reduced reservation in the Dinehtah.

The Navajo were, and remain, different from their Indian neighbors. Unlike the cliff-dwelling Anasazi (the Old Ones) and their Hopi and Pueblo descendants, the Navajo have traditionally been herders who live in small family compounds made up of hogans, small dome structures that stand as microcosms of the harmonious way of life, the "beautiful rainbow" way of which the Navajos often speak.

The mythology of the Dine contains elements borrowed from their neighbors, whose cultures had been established in the Southwest for centuries before the Navajo arrived. It is a complex mythology associated with rituals. There are inevitably several variations of a given Navajo myth, because, unlike the Hopi and other Pueblo people, they have not tended to live in the large concentrated village groups that are more conducive to orthodoxy in myth and ritual. The creation myth itself exists in several versions as sung by various sacred singers or shamans. Still, there is a basic consistency to Dine mythology that transcends the details. The myths are passed on orally from old teacher-shamans to younger ones and are treated with the respect reserved for sacred texts in any culture. The Dine myths are a "history" of the people and an expression of the right way, much as the Old Testament, with all its variations and contradictions, is a history of the Jews and a reflection of the Jewish approach to being.

It should be understood that the creation story, the very center of Navajo mythology, is meant to accompany particular rituals, especially curing rituals and other rites of new beginning. To hear the whole Dine creation, one would have to attend a Blessing Way, the nine-day ceremony that is the most holy of Navajo ceremonies. In the Blessing Way, the creation is not only chanted: it is reflected microcosmically in the hogan in which the ceremony takes place and visually re-created in elaborate but temporary sand paintings, the colors, directions, and figures of which represent the basic principles on which the universe is based. The Navajos have no temples or

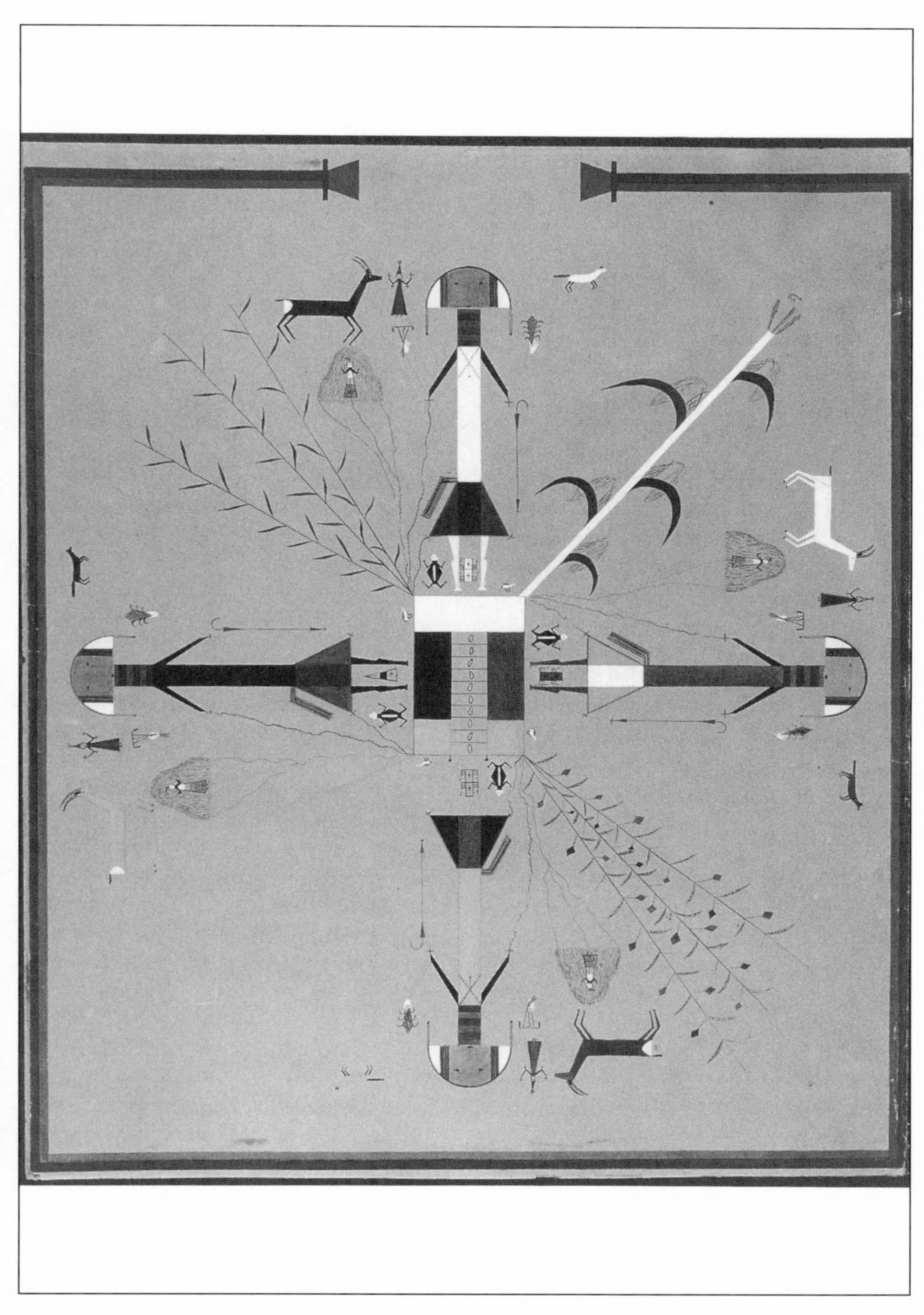

The traditions of the Dine, or Navajo, of the American Southwest seem to date from about the tenth century. Elaborate paintings using sands of different colors accompany ceremonies that include the Blessing Way, an account of creation that takes nine days in the telling; The Emergence *is a painting from the Blessing Way.*

kivas (underground religious spaces of the Pueblo people), but the hogan and the sand painting are more than adequate as sacred and symbolic space.

The Navajo creation is an emergence myth (see also **Creation by Emergence**). It begins with a dark first world, or, as it turns out, underworld. Above that first world was another domed world, and another, and still another before our world, the fifth world. The people made their way through these worlds, evolving by the fourth world into people as we know them. Above our present world there is thought to be a sixth world of perfect harmony.

The First World, an island floating in the endless oceans, domed by the hoganlike sky, and secured by the four directions, was populated by the Insect people. They really were more like insects than what we think of as people. Morning for them was when white came up in the east and day was blue in the south. Evening was yellow in the west, and night was black in the north. The gods of the Insect people lived in the surrounding seas—these were Water Monster in the east, Blue Heron in the south, Frog in the west, and White Mountain Thunder in the north.

The Insect people were quarrelsome. They recognized no sexual taboos and paid no attention to the warnings of their gods. The gods told them to leave the First World and forced them to do so by creating a great flood, which covered the First World (see also **The Flood and Flood Hero**). The Insect people flew up to the hard sky; through a hole in the east they entered the Second World, which was blue.

The Second World was populated by the Swallow people, who lived in strange rough houses with holes in their tops. After searching the world in vain for people like themselves, the Insect people made friends with the Swallow people and lived with them. One day an Insect sexually assaulted a Swallow, however, and once again the Insects were expelled from a world. Led by the Locust, they flew to the top of the Second World, until Nilch'i, the wind, appeared and told them how to find in the south the entrance to the Third World, which was yellow.

In the Third World, the Insect people discovered the Grasshopper people, who lived in holes in the ground. After searching in vain for people like themselves, they befriended the Grasshoppers and lived with them. As in the Second World, however, they upset the harmony by way of sexual assault and once again they were ordered to leave. They flew up to the sky, and after searching for an exit, they were led to an opening in the west.

The Fourth World was black and white. There was no sun, moon, or stars, no real day. There were four mountains on the horizons of each of the directions. There did not seem to be inhabitants, but Locust scouts reported that they had in fact found people in the north who lived in houses and who grew things in fields. These were the Pueblo people, the Kisani, who visited the Insect people and offered them kindness and food. The Insect people vowed among themselves not to make the mistake here that they had made in the other worlds, and things went well for quite a time.

Then they were visited for four days in succession by four strange beings who made incomprehensible signs. These were White Body, Blue Body, Yellow Body, and Black Body. On the fourth day, Black Body explained to the Insect people that the gods wished to make more people, but ones that looked like themselves, without the odd appendages and foul smells of the Insect people. Black Body instructed the Insects to bathe themselves and await the return of the gods in twelve days.

The Insect people bathed themselves on the morning of the twelfth day and dried themselves with corn meal—white for males, yellow for females. Then the gods called out as they approached, appearing on the fourth call. Blue Body and Black Body carried buckskin, while White Body and Yellow Body carried one yellow and one white ear of corn. The gods performed a sacred ceremony during which supernatural Mirage people appeared, the yellow ear of corn became a woman, and the white ear of corn became a man. The wind came and gave breath to these beings. The gods instructed the Insects to make a brush hogan for First Man and First Woman, who lived together as husband and wife.

After four days, First Woman gave birth to hermaphrodite twins. After four more days she produced another set, a boy and a girl. Five sets of twins in all were born, and in each case they became fully grown in four days and, except for the hermaphrodites, each set lived together as husband and wife.

First Man, First Woman, and each set of twins were taken at four-day intervals to the home of the gods in the east and were taught the mysteries of life, including witchcraft. They were taught how to wear masks in certain ceremonies and how to pray for necessities. They learned to impersonate the talking god, the house god, and many others. They also learned to keep their incestuous marriages secret, to marry outside of their own immediate family, and to marry among the Insect people. Soon they inhabited the land. They built dams and planted fields, like the Kisani. One of the hermaphrodite twins invented pottery, the other the wicker water bottle. The People also learned to hunt, and with the help of the gods they learned to use deer heads as masks to make the hunt more successful.

One day the sky (Father Sky) and the earth (Mother Earth) seemed to slam together, and when they separated (see also **Creation from Division of Primordial Unity**) Coyote and Badger appeared. Coyote hung about the People's farm, and Badger went into the hole to the lower world.

The Kisani chief taught the People about the four sacred mountains—Mount Blanca in the east, Mount Taylor in the south, the San Francisco Peaks in the west, and Mount Hesperus (Navajo Mountain) in the north.

Then the People began to quarrel. First Woman accused First Man of being lazy and of only working in order to have the pleasure of her vagina. "We do not need you men," said First Woman. So it was that the men and women separated and lived on opposite sides of the river. The men persuaded the Kisani people—men and women—to come to their side, and they took all their tools with them. During the first winter both sides did well. The men worked hard to provide for themselves in their new place, and the women lived on the stores of the old settlement. The women sometimes came to the river and teased the men. Over the next years, the fields of the men did well, and the women slacked off. Soon the women were starving and they no longer taunted the men across the river. Many tried to swim across and were taken by Water Monster. Finally, the men allowed the women to cross over, and after cleansing ceremonies the men and women came together again. Two young girls were taken by Water Monster on the way over.

With the help of the gods and their mysterious ceremonies, the People went under the waters and found the lost children with Water Monster and two of his own children in the North room of many colors. The People were allowed to take their children, but **Coyote,** who had descended with them, stole the children of Water Monster (see also **Trickster**). Since Coyote always kept his robe closely wrapped around him, he was able to conceal his theft from the People, but the wrath of Water Monster was not hidden. Soon a great flood came. The people were protected from the flood only at the last minute. A strange old man and his son appeared and hid them in a huge reed. Since Turkey was at the bottom of the reed, his tail feathers got wet at the tip and turned white, as they remain today.

The People sent scouts to the sky to seek a means of escape from the Fourth World. It was Locust who finally succeeded in digging through to the Fifth World, but the water followed the People up the reed and began to enter the Fifth World, too. Only when the children of Water Monster were discovered and thrown down the hole did the waters subside.

After the flood, the People looked about for food. The Kisani had brought some corn kernels, and some of the People threatened to take the new corn. Eventually the Kisani, the Pueblo people, and the People (the Dine) went their own ways.

First Man, First Woman, Black Body, and Blue Body then made the world of Dinehtah. They reestablished the sacred mountains, made male rain (hard rain) and female rain (soft rain), and brought up the gods from the Fourth World. Finally, they made the sun and moon, and there was light in the Fifth World.

This is a version of the Fifth World emergence as told to Old Man Buffalo Grass by his grandmother, Esdan Hosh Kige.

The Fifth World

First Man was not satisfied with the Fourth World. It was a small, barren land; and the great water had soaked the earth and made the sowing of seeds impossible. He planted the big Female Reed and it grew up to the vaulted roof of this Fourth World. First Man sent the newcomer, the badger, up inside the reed, but before he reached the upper world water began to drip, so he returned and said he was frightened.

At this time there came another strange being. First Man asked him where he had been formed, and he told him that he had come from the Earth itself. This was the locust. He said that it was now his turn to do something, and he offered to climb up the reed.

The locust made a headband of a little reed, and on his forehead he crossed two arrows. These arrows were dressed with yellow tail feathers. With this sacred headdress and the help of all the Holy Beings the locust climbed up to the Fifth World. He dug his way through the reed as he digs in the earth now. He then pushed through mud until he came to water. When he emerged he saw a black water bird swimming toward him. He had arrows crossed on the back of his head and big eyes.

The bird said: "What are you doing here? This is not your country." And continuing, he told the locust that unless he could make magic he would not allow him to remain.

The black water bird drew an arrow from back of his head, and shoving it into his mouth drew it out his nether extremity. He inserted it underneath his body and drew it out of his mouth.

"That is nothing," said the locust. He took the arrows from his headband and pulled them both ways through his body, between his shell and his heart. The bird believed that the locust possessed great medicine, and he swam away to the East, taking the water with him.

Then came the blue water bird from the South, and the yellow water bird from the West, and the white water bird from the North, and everything happened as before. The locust performed the magic with his arrows; and when the last water bird had gone he found himself sitting on land.

The locust returned to the lower world and told the people that the beings above had strong medicine, and that he had had great difficulty getting the best of them.

Now two dark clouds and two white clouds rose, and this meant that two nights and two days had passed, for there was still no sun. First Man again sent the badger to the upper world, and he returned covered with mud, terrible mud. First Man gathered chips of turquoise which he offered to the five Chiefs of the Winds who lived in the uppermost world of all. They were pleased with the gift, and they sent down the winds and dried the Fifth World.

First Man and his people saw four dark clouds and four white clouds pass, and then they sent the badger up the reed. This time when the badger returned he said that he had come out on solid earth. So First Man and First Woman led the people to the Fifth World, which some call the Many Colored Earth and some the Changeable Earth. They emerged through a lake surrounded by four mountains. The water bubbles in this lake when anyone goes near.

Sources: Locke, 55–79; Zolbrod.

Nebular Creation Theory

According to this late eighteenth-century theory of French scientist Pierre Simon de La Place, the solar system developed from a huge nebula that rotated at great speed during a long cooling process and literally threw off the planets (see also **Creation in Science**).

Source: Leach, 17–18.

Negritos Creation

The Negritos, a pygmy people of Malaysia, have an *ex nihilo* creation myth characterized by **creation by thought** or dreaming (see also **Creation from Nothing; The Dreaming**).

In the beginning there was only the divine couple, Pedn and Manoid, and there was the sun, but no earth. It was the dung beetle who created the earth out of mud (see also **Creation from Clay**). After a while, the couple came down to the new earth. They had children, but in a strange way. It came about that, in a dream, Manoid begged Pedn for a child, and Pedn went out and picked some fruit from a tree, which became a boy child when he placed it on a cloth. When Manoid dreamed of another child, Pedn created a girl child on the cloth with the seed from the tree. The boy was Kakuh-bird and the girl was Tortoise. They married and had children, one of whom shot an arrow into a rock that then gave forth water.

Source: Long, 210–211.

Netsilik Creation

This is the *ex nihilo* creation myth of the Netsilik (Seal people) of Greenland (see also **Creation from Nothing**).

In the beginning there was only darkness, and the animals and people were the same and interchangeable. They all talked and lived the same way. Words were the most powerful things in those days (see also **Creation by Word**). It was the hare who created day by saying, "Day."

There are many stories of the beginning. Some say that after a great flood (see also **The Flood and Flood Hero**) there were two men left; one of them, a shaman, made the other into a woman. She had children by him, and this is how the people began.

What follows is part of the Netsilik myth as told by the wise woman Nalungiap.

> In those times there were no animals in the sea; people knew nothing of burning blubber in their lamps. At that time newly drifted snow would burn, the soft, chalky-white heaps of very fine snow that gather in the shelter of the firm, hard drifts, the kind we call apᴇrʟɔrqᴀ˙q. No one needed blubber then. This story, people say, is a distant memory of the very first days, the time when the first people lived on earth and had to travel far from one place to another. For

they had to go far to get something to live on. But at that time they knew magic words that could move houses; they just sat still in their houses and said magic words, and then they rushed through the air with house and everything in it, to a new settlement where they started at once to break up the ground to find food.

That was the time people lived in darkness, in the very first beginning, when there were only men and no women. Then forests grew on the bottom of the sea, and it is the remains of those forests that to this day tear themselves loose when the storms blow, so that we find driftwood on our shores.

And in the first days there was no ice on the sea. The sea was always open and never closed by ice. Sea ice came from an angry old witch-woman. She wanted to kill a man. The man was Kivioq, the one you know from the story. The witch was angry because she had not got him to eat, and when he got down into his kayak she threw her ulo at him; it made "ducks and drakes" over the water and turned to ice. That is where the sea ice came from, the story says.

Woman was made by man. It is an old, old story, difficult to understand. They say that the world collapsed, the earth was destroyed, that great showers of rain flooded the land. All the animals died, and there were only two men left. They lived together. They married, as there was nobody else, and at last one of them became with child. They were great shamans, and when the one was going to bear a child they made his penis over again so that he became a woman, and she had a child. They say it is from the shaman that woman came.

That is all I know about people. I have also heard that the earth was here before the people, and that the very first people came out of the ground from tussocks. But these are hard things to understand, difficult things to talk about, all this about where something began, where the first people came from. It is sufficient for us to see that they are here and that we ourselves are here.

And there are those who say that the children of the earth were not the first people, and that they only came to make people many. Women who happened to be out wandering found them sprawling in the tussocks and took them and nursed them; in that way people became numerous.

And the earth. Here we only know our land. It has become habitable because the Tunrit first came here and found out how to hunt the game. But we know that our land is not the whole earth, for the earth has no boundaries, and he who wants to can keep on travelling on and on. The earth was as it is at the time when our people began to remember.

Source: Weigle, 231–237.

New Guinea (Papua) Creation

According to their creation myth—one that has emergence aspects—Gainji was the creator of the Papuan Keraki tribe of New Guinea in Melanesia (see also **Creation by Emergence**). They say the first humans came out of a palm tree. It was Gainji who heard the people talking in their many languages in the palm. When they were out, the people went away in their own language groups, and so it is today.

The Papua Kiwai people say that Marunogere was the creator. He gave people the pig and taught them about the sacred ceremonies.
Source: Leach, 175–176.

New Hebrides Creation

In the Melanesian New Hebrides Islands, there is a legend about two brother-twins, one wise, one foolish. To preserve the wise brother's creative work, the foolish brother had to be exiled from the world.
Source: Leach, 176.

Ngombe Creation

The *ex nihilo* creator god of the Ngombe people of the wilds of Zaire is Akongo, the Mysterious One, the source of being (see also **Creation**

from Nothing). This is a story of the origins of evil, one that brings together the familiar figures of the corrupted woman and the monster (see also **Hebrew Creation Myths**). In the beginning Akongo lived with us in the sky, but he got so tired of human quarrels that he left and has not been seen since. Some people tell a more complicated tale, however, and this is how it goes:

The woman Mbokomu was so irritating to Akongo and the people that Akongo dropped her, her children, and some food down to earth in a basket. The family planted and did well on earth, but Mbokomu was afraid her family would die out. She convinced her son to take his sister as a wife. He did so, unwillingly, and his sister became pregnant.

The sister was walking about one day when she met a very hairy but pleasant creature whom she grew to like. His name was Ebenga, and when the sister shaved him he looked quite like a man. Ebenga was evil, however, and he put a curse on the woman and her child. The child was born and grew up to be a witch. He plagued all of the other children born to the brother and sister, as well as their descendants (see also **Incest in Creation**).
Source: Sproul, 47–48.

Ngurunderi Creation

The Ngurunderis are aboriginals living in the Lower Murray River area of southern Australia. As with virtually all aboriginals, their creation myth is a Dreaming (see also **The Dreaming**), the description of an ancient creative "walkabout" by sacred ancestors, which perhaps signifies a kind of *ex nihilo* creation (see also **Creation from Nothing**).

The great ancestor Ngurunderi canoed down the Murray River in search of his two runaway wives. A giant fish swam ahead of the ancestor, creating the present river out of the tiny stream that it used to be. When Ngurunderi tried to spear the fish, he missed, but the spear became Lenteilin, the Long Island. Later, when the ancestor succeeded in spearing the fish, he cut it up, forming all the different fish the people find today.

Ngurunderi set up camp, but when he sensed the nearness of his wives, he left everything and went in pursuit of them; his canoe became the Milky Way. The wives heard their husband coming and abandoned their own camp in favor of a reed and grass raft, which turned into the reeds and grass trees of the area where they left it on the far side of Lake Albert.

In Kingston, still following his wives, the ancestor was challenged by the sorcerer, Parampari. Victorious after a long fight, Ngurunderi burnt his adversary's body, which remains there today as the great granite boulders.

Ngurunderi continued along, creating islands with his spears and places like the Longkuwar Bluff with his club.

The ancestor caught up with his wives on Granite Island, but they ran to Cape Jervis, from which point they ran toward the then connected bit of land now called Kangaroo Island. That island was created when Ngurunderi, in anger, caused waves to sweep across the connecting land, drowning not only the land but his wives. They became the little Rocky Pages Islands. The Dreaming ended, and Ngurunderi rose into the Milky Way.
Source: Berndt, 164–185.

Nigerian Creation

See Efik Creation; Ijaw Creation; Nup Creation; Yoruba Creation.

Nihongi

The Nihongi, or Chronicles of Japan, was written in 720 C.E. It is the major source for the **Japanese creation** story and other myths.

Norse Creation

There is, in addition to the Icelandic myth, another Norse creation myth (perhaps from Northern Germany) that says the sun and moon came from the sparks of the fire world, Muspelheim. The sparks were formed into Maane (moon) and Sol (sun) by their father, Mundilfare. The gods were angry at Mundilfare's presumption, however, and they exiled the children to the sky (see also **Icelandic Creation**).
Source: Olcott, 25.

Nugumuit Creation

The Nugumuit are Eskimo people of the Frobisher Bay area who believe in a high father god named Anguta, who made everything that is (see also **Eskimo Creation**).
Source: Leach, 47–50.

Nup Creation

The Nup people of Nigeria tell an *ex nihilo* creation story, which, like so many others, tries to explain the mystery of death. It tells us, in effect, that we living things are death-defined, that death is a necessary aspect of the fertile life (see also **Creation from Nothing**).

First God created tortoises, humans, and stones. He made males and females of each of these, but they could not reproduce. When they got old they just became young again. The creatures wanted children, however. First the tortoises asked for children. God told them that with children would also come death, but the tortoises insisted. They had children, and then death began for them. The humans made the same demand. Finally God gave in, and children and death came to them as well. The stones learned from what they saw happening to the others, so they did not ask God for children, and they received neither children nor death.

Source: Beier, 58–59.

Nyamwezi Creation

In Tanzania the Nyamwezi tell how the world was created by Shida Matunda and undermined by one of the first women.

After he created the world *ex nihilo,* Shida Matunda made two wives for himself (see also **Creation from Nothing**). It is said that after the death of the first wife, the god's favorite, he buried her in her hut, watching over the grave and watering it regularly. Shida Matunda saw a beautiful little plant growing from the grave. He knew now that his dead wife would be reborn.

The second wife, whom Shida Matunda had ordered to stay away from the grave, was jealous, and she stole into the hut and cut down the plant. Blood poured out and filled up the hut. "What have you done, Woman?" the god cried. "You have prevented the rebirth of the first wife, and so you have brought death into the world." So began the way things are.

Source: Beier, 62.

Okanagan Creation

The Okanagan are an ancient tribe of the Pacific Northwest. Theirs is an animistic creation-from-formlessness myth in which earth itself is the primal being (see also **Animism; Creation from Chaos**). In its present form, the myth is clearly influenced by white Christian settlers.

Old One created a woman as the world. He took the tiny earth that was first here, rolled it out, and pulled it like dough until it became Earth Woman. Her head was in the west, where the rivers flow and where we go when we die. Trees and other plants are her hair, the soil is her body, stones are her bones, and the wind is her breath. Old One made the animals out of little pieces of her body. He blew on them to give them life (see also **Creation from Clay**).

The Indians, made of red clay, were among these new creatures. Some of them were wicked, so Old One sent his son—some say Jesus—to put things in order. The people killed the son, however, and he went back up to Old One.

Old One sent down **Coyote** to try to put things to right (see also **Trickster**). He was more successful—he killed monsters, taught the people things they needed to know, and divided them into tribes with different languages. He also made several foolish mistakes, however, so Old One took the form of a wise old man and decided to complete the work of creation himself. He met Coyote on the road, but Coyote refused to recognize him when he said he was Old One, chief of the world. Only when Old One picked up a river and moved it did Coyote recognize him. Then Old One sent Coyote away, thanking him for the work he had done.

Finally, Old One taught the people how to pray so they could talk to him and he to them after he returned to the sky. He told the people that in time Earth Woman would be ancient enough for Old One and the dead to return to her. Then everyone would live together and revere Earth Woman as the Great Mother (see also **Mother Creatrix**).
Source: Leach, 55–57.

Old Man as Creator

The name Old Man is frequently applied to the North American Indian trickster-creator. Sometimes Old Man (Old One or Wise One) is Raven, sometimes Coyote. Sometimes, as among the Tananas, he is Beaver Man (see also **Blackfoot Creation; Crow Indian Creation**).

Omaha Creation

The Omaha Indians of the American Midwest see creation as beginning *ex nihilo* and volcanically in the mind of the Great Spirit (see also **Creation by Thought; Creation from Nothing**).

In the beginning there was only Wakonda, the Great Spirit, and all things—plants, animals, and humans—were spirits in his mind. These beings wanted to take form, but they could only wander about in the space of God's mind, which is the space between heaven and earth. They tried to live on the sun and then on the moon, but these places were not satisfactory. Finally they descended to earth, where they found only water. Journeys in the four directions revealed only more water. Then suddenly a great rock emerged from the center and burst into fire, sending water into the air as clouds and leaving some land. Now the creatures had a place to live, and they became real. They were grateful to Wakonda, and they praised him.
Source: Eliade (B), 84.

Onondaga Creation

The Onondaga are an Iroquoian tribe (see also **Iroquoian Creation**). Their creation myth has developed over the centuries, and continues to develop, since creation myths must take into account the way things are as well as the way they once may have been. There are elements in this myth of the earth-diver and star woman motifs so common to other

Northeastern Indians (see also **Cherokee Creation; Earth-Diver Creation, Huron Creation**).

They say that there were once man-beings who lived in the sky of the world above this one, and that a woman-being went there with a comb and began straightening out the hair of one of the man-beings. Soon she became pregnant and the man-being became the first to experience the mystery of death, for with birth must come death. The man-being was placed by his mother in a coffin.

When the woman-being gave birth to a girl, her mother (the Ancient One) asked the woman-being who would be the child's father, but the woman-being did not answer. The child grew, and one day she began crying and would not stop. It was the Ancient One who told her daughter to take the child to the male-being's coffin. When the child saw the coffin she was happy. The corpse of the man-being gave her instructions on the right way to be until she married.

When the girl child herself had a baby called Zephyrs, her husband, a chief, became ill. He sang a song, telling the other man-beings to pull the tree called Tooth that grew near his hut. Through the hole left by the tree, he threw his wife and Zephyrs down to the world below—our world.

The woman-being, Star Woman, fell and fell and saw only water beneath her. The animals below saw her falling and decided to make land for her. Many animals tried to dive below the waters to get earth, but only Muskrat succeeded, and he died in the process. With the earth he brought up, however, the animals made land on the Turtle's back. Then the flying creatures formed themselves into a huge net in which they caught the falling woman, and they brought her safely to the new earth.

Star Woman and her daughter brought fire and taught the people the art of hunting. When the daughter had grown she was visited in the night and she soon became pregnant. Just before she gave birth, she heard two male-beings talking inside her body, arguing about how to be born. One came out by the normal way, the other by an armpit. This armpit child killed his mother and told his grandmother the other son had done the deed. Thus, there are good and evil people. These were the first man-beings on earth.

Source: Weigle, 194–202.

Oqomiut Creation

The Oqomiut, like many of the Central Eskimos, tell the myth of a woman, Sedna, who is the primary force in the creation of the world's creatures. As for the world itself, it is only known that it was created *ex nihilo* by Anguta (see also **Creation from Nothing; Nugumiut Creation**).

Once there was Anguta, who lived with his daughter, the beautiful and much-desired Sedna. When a great seabird, a fulmar, flew over her one spring day and urged her to follow him over the sea to his home, Sedna went. When she got there, however, she was horrified by the fulmar's foul tent and lack of food. She called in despair for her father, and after a year had passed, he came with the warm winds that broke up the ice. He killed the fulmar and put his daughter in his boat for the journey home.

When the other fulmars found the body of their chief they mourned—as they still do today with their sad cries—and they became angry. They searched the sea for the murderer and blew up a huge storm when they saw the boat of Sedna's father. To save himself, the father threw his daughter overboard and chopped her fingers when she tried to hold onto the boat. Parts of the fingers fell into the water and became whales; other parts became other fish and sea creatures. Finally Sedna fell into the sea and the storm subsided. Sedna was not dead, however, and she climbed back into the boat.

Sedna did not love her father any longer, and when he was asleep, she ordered her dogs to bite off his hands and feet. This they did before he woke up. In fury, Anguta cursed everything and everyone, and the earth swallowed him, the dogs, and Sedna. Just before going down, Sedna created the deer.

Sedna, Anguta, and the dogs now live under the world in Adlivun. Sedna rules there, and Anguta hobbles around there with no feet. When people die, they go to Sedna's house in Adlivun. The bad ones have to sleep next to Anguta, who pinches them.

Source: Leach, 47–50.

Origin Myths

Origin myths are sometimes considered to be the same as creation myths, but more accurately, they are myths about how specific aspects of existence came about. "How the Zebra Got Its Stripes" and other such anecdotes would be contained in origin myths.

Orphic Creation

The Orphic cult developed in Greece beginning in the late seventh century B.C.E. Many of the characters in the Orphic myths are taken from the related, mainstream Olympian religion (see also **Greek Creation**), but Orphism also includes a mystical element, an emphasis on the soul's path to salvation, and such Eastern principles as reincarnation. The Orphic creation story is an example of **creation from a cosmic egg.** It stresses the androgynous aspect of the creator, androgyny being the symbol of absolute wholeness and perfection, that state toward which the Orphics strove in order to free themselves from the restrictions of the dualistic, gender-based (and, therefore, fertility-centered, death-defined) life of the body.

In the beginning was the silver cosmic egg, created by Time. Phanes-Dionysos broke forth from the egg as the firstborn (Protogonos), the androgynous container of all the seeds of life. It was Phanes-Dionysos who created the universe, beginning with a daughter, Nyx (Night), and later the familiar gods, Gaia and Uranos.

It was also said that when Zeus came along, he swallowed Phanes-Dionysos and thus, by containing the source of being in his own belly, made the world new.

The Orphics also told of the god Dionysos being eaten by the Titans. When the Titans were destroyed, mankind emerged from their ashes. Mankind therefore contains the evil of the Titans and the goodness of Dionysos.

Sources: Long, 117–120; Sproul, 169.

Osage Creation

The Osage were a Midwest people who eventually migrated to the Great Plains of North America. They are children of the sun and moon, and their myth features the familiar **earth-diver creation** figure of the fall from heaven.

In the beginning the People lived in the sky. When they asked the sun and moon who their parents were, the sun said he was their father, and the moon said she was their mother. The moon said it was time for them to go down to the earth. They did this, but there was only water there, so they could not land. They floated in the air calling for help, but no one came.

Finally the elk, one of the animals floating down to earth with the people, came to everyone's assistance by letting himself fall into the water. As he sank he called on the winds to blow away the waters until they flew off as mist, leaving land. The elk rolled in the first mud that appeared, and from his loose hairs that remained in the soil, all the plants and trees grew.

There are some Osage who say that the first people came from a union between two animals, Beaver Girl and Snail Boy. These Osage say that

Beaver Girl and Snail Boy produced a son and a daughter that were neither snail nor beaver, but Osage. These were the first people, naturally enough, and they made houses that look like beaver houses (see also **Birth as Creation Metaphor**).
Sources: Erdoes, 119; Marriott (B), 21–25.

Pandora

Before Hesiod told the patriarchal story of Pandora's box, from which Pandora is said to have released the woes that still afflict humankind, Pandora was known as an incarnation of earth, the creative Earth Goddess whose name means gift-giver. She is surely related to an earlier **mother-creatrix.**

Papago Creation

The Papago Indians of southern Arizona say that the world was created by Elder Brother, helped by **Coyote** (see also **Trickster**). The Papago creation myth is told only at night in winter when the snakes are asleep, since they would be offended at the revelation of the mysteries if they were to hear it. Only certain people can tell the Papago creation myth. In each village there is an old man who sings and tells the sacred story while the men smoke a ceremonial cigarette. During the recitation, no movement or noise can be made or the story-telling must stop.

Some parts of the creation story are known to many people. Children, for example, know the story of the woman of ancient times who gives birth to the twins in the mountains. The twins grow immediately, and their mother sends them to the nest of the eagle to get arrow feathers. Then she shows them how to make arrows out of wood. At each stage the woman sings into life whatever has been made: "Now bows are made/Arrows are made/Into the West they go/Watch them fly, my boys."

The mother sends the twins for canes with which to make flutes, and they become the flute players who lure and marry the daughters of Buzzard, the desert bird. Buzzard's companion, Blue Hawk, kills the twins and sings of his own power to destroy.

One of the girls, who has become pregnant, gives birth to a hero who takes revenge by killing Buzzard, singing, "Joy comes to the man who kills his enemy." He takes the scalp of Buzzard to the grandmother, and they go off together to the far country. The grandmother and the hero are pursued by the now angry Buzzard girls, who sing the song of crazy women with painted faces. The hero makes a bridge across the ocean with his bow, which is the rainbow. He and the grandmother cross the bridge and then twist it to throw off the vengeful women. They fall into the sea and turn into birds. Now the grandmother and the hero sing a song of praise.

The following anecdote is Papago Indian Ruth Underhill's childhood memory of the circumstances around his father's recounting of the creation story.

> I knew all about Coyote and the things he can do, because my father told us the stories about how the world began and how Coyote helped our Creator, Elder Brother, to set things in order. Only some men know these stories, but my father was one of them. On winter nights, when we had finished our gruel or rabbit stew and lay back on our mats, my brothers would say to my father: "My father, tell us something."
>
> My father would lie quietly upon his mat with my mother beside him and the baby between them. At last he would start slowly to tell us about how the world began. This is a story that can be told only in winter when there are no snakes about, for if the snakes heard they could crawl in and bite you. But in winter when snakes are asleep we tell these things. Our story about the world is full of songs, and when the neighbors heard my father singing they would open our door and step in over the high threshold. Family by family they came, and we made a big fire and kept the door shut against the cold night. When my father finished a sentence we would all say the last word after him. If anyone went to sleep he would stop. He would not speak any more. But we did not go to sleep.

Sources: Bierhorst, 95–96; Weigle, 170.

Papua Creation

See New Guinea Creation.

Pawnee Creation

The Pawnees are hunters of the Nebraska and Kansas plains. Their *ex nihilo* creation story is traditionally told during the spring renewal ceremonies (see also **Creation from Nothing**). At this time the bundle of the holy Star Woman, Yellow Buffalo Calf, is opened. It contains ears of corn and the sacred pipe. During the ceremonies, the priests praise Mother Corn and recreate the world through a holy dance and a singing of the creation story. Long ago, the Skidi Pawnee would sacrifice a boy or a girl to the Morning Star as part of the holy spring rites.

In the beginning was space itself, Tirawahat (Tirawa). It was he who organized the gods in creation. He placed Sun in the east and Moon in the west. Evening Star was to be the mother and was placed in the west. Morning Star was the brave in the east who would chase the star people into the west. Then Tirawahat made four other stars the supporters of the four corners of the world. He gave the wind, the thunder, the lightning, and the clouds to Evening Star, and they sang, rattled, and danced as Tirawahat created the earth.

He dropped a pebble into the clouds and there was water. The earth-supporters struck the water with their clubs and earth was formed. To populate the earth, Evening Star took Morning Star as her husband, and they produced Mother of Humanity. Sun and Moon produced Father of Humanity. Evening Star then made the sacred bundle, and the elements—clouds, wind, thunder, and lightning—taught the new people the sacred songs and dances they still do, calling it the Thunder Ceremony.

Some say the elements became Paruhti, who crosses over the barren winter earth and brings the new life of spring.

Still others, who mistrust the power of women, say Evening Star, or West Star Woman, had to be overpowered by Morning Star, the Great Star, before creation could be accomplished. Followed by his brother, who carried the sacred bundle, Morning Star headed west toward Evening Star,

who moved along luring Morning Star toward her. When Morning Star got close to her, however, Evening Star placed things in his way to hinder and tease him. For instance, once she beckoned to him, and as he approached she caused the earth to open and all the waters of the heavens to fall down into the mouth of the great serpent. Morning Star won out by throwing a ball of fire into the serpent's mouth. When he got to the lodge of Evening Star, he had to overcome the stars of the four directions. Then he had his way with Evening Star.
Sources: Bierhorst, 167–168; Williamson, 222.

Pelasgian Creation

The Pelasgians arrived in Greece long before the more patriarchal originators of the Olympian religion; the Olympian creation myth was influenced by that of the Pelasgians (see also **Gaia; Greek Creation**). The Pelasgians were goddess worshippers. Their creation myth is dominated by a female creatrix and a primal cosmic egg (see also **Creation from a Cosmic Egg**). In the myth we find a female-serpent relationship that reminds us of the Hebrew creation, but the emphasis is on female power rather than weakness.

In the beginning there was the great goddess Eurynome, who emerged naked from chaos and divided the waters from the sky so she could dance lonely upon the waves. As she danced she created the wind (see also **Creation from Nothing**). She caught the north wind and rubbed it, and it became the serpent, Ophion. Ophion coupled with the dancing goddess and she was full.

Now, as a dove, Eurynome laid the world egg, and she ordered Ophion to encircle it until it hatched the sun, moon, stars, and earth with all of its creatures and plants.

The goddess and her companion lived on Olympus until Ophion became arrogant and had to be banished—his head flattened and his teeth broken—to the darkness under the earth.

Later Eurynome made the Titanesses and Titans to control the planets, and finally she made the first man, Pelasgus.
Source: Graves (Vol. 1), 27.

Persian (Iranian) Creation

The mythology of early Iran was repressed by the emergence of Zoroastrianism in the sixth century B.C.E. (see also **Zoroastrian Creation**).

The early mythology was probably influenced by the mythology of Sumer and bears a great deal of resemblance to the Vedic mythology of India in about 1000 B.C.E. For example, Yima, a solar deity (who becomes the Zoroastrian Jamshid), is similar in name to the Vedic Yama.

Yima was a god of fertility who used a golden arrow to pierce the earth and thus make it pregnant. He then caressed the earth with a golden scourge and it doubled in size.

Yima marries his sister and is punished for this sin by being killed by the serpent, Ahzi Dahak.

Source: Freund, 155–156.

Philippine Creation

See Bagobo Creation.

Pima Indian Creation

The creation myths of the Pima Indians of southern Arizona share many characters and situations with the myths of their neighbors, the Papago (see also **Papago Creation**) and the Yuma (see also **Yuma Creation**). The Piman creation, like the Yuman creation, features a dying god figure. It also makes use of the emergence theme (see also **Creation by Emergence**).

At the beginning of time Earthmaker (some say it was Great Magician) made the sky and the earth, who mated and produced Elder Brother. Elder Brother, assisted by Coyote, created humanity after an earlier people was washed away by a great flood (see also **The Flood and Flood Hero**).

There are many stories of this human creation. One says that Earthmaker's humans were badly made, and that Earthmaker descended into the earth after an argument over this with Elder Brother. According to this version, Elder Brother created the Hohokam people after Earthmaker's death, but he disrupted their harmony by molesting maidens during the puberty ceremony. He was killed for his sin either by the people themselves or by Buzzard.

Elder Brother came back to life and went with the sun into the underworld and brought up the Piman people, who defeated the Hohokam in battle. Before battles, the Pimans always told the story of Elder Brother's revival and his leading an army out of the underworld.

Some of the Pimans say it was Earthmaker who, with Coyote's misguided assistance (see also **Trickster**) created humankind (see also **Imperfect

Creation). According to this version, on his first try Earthmaker modeled a little man out of clay and put it in the oven, but **Coyote** changed its shape and it became a dog. On the second try Earthmaker made two figures, but Coyote urged him to take them out of the oven too soon and they were underdone. These were sent away as white people to other lands. The third batch Earthmaker overcooked on Coyote's advice, and the people came out dark brown. They were sent across the ocean to another land. For the final batch Earthmaker ignored Coyote's advice and the people came out perfectly toned. These, of course, were the Pimas, and Earthmaker was satisfied.
Sources: Bierhorst, 103–104; Erdoes, 46–47.

Poetic (Elder) Edda

With the **Prose Edda** of Snorri Sturluson, the Elder Edda is the source for early Scandinavian/Icelandic mythology (see also **Icelandic Creation**). It was written down between the ninth and twelfth centuries C.E. and discovered in 1643. It is sometimes called the Edda of Saemund, because it was wrongly attributed to Saemund Sigfusson, a writer of the twelfth century.

Pohonichi Miwok Creation

These Native Californians tell an **earth-diver creation** that features **Coyote** (see also **Trickster**) and is related to the **Yokut creation** and many others like it.

Once there was only water and Coyote. Coyote sent some ducks to look for earth. One of the ducks succeeded, and Coyote took the soil from its beak. Then he sent the duck down for seeds. Coyote mixed the soil, seeds, and water, and the mixture grew until it became the earth.
Source: Long, 208.

Polynesian Creation

The most common of the Polynesian creation myths—of which there are several, with many variations of plot and names—describes the god Tane as an artist-creator (see also **Creation by *Deus Faber***). It was Tane, for

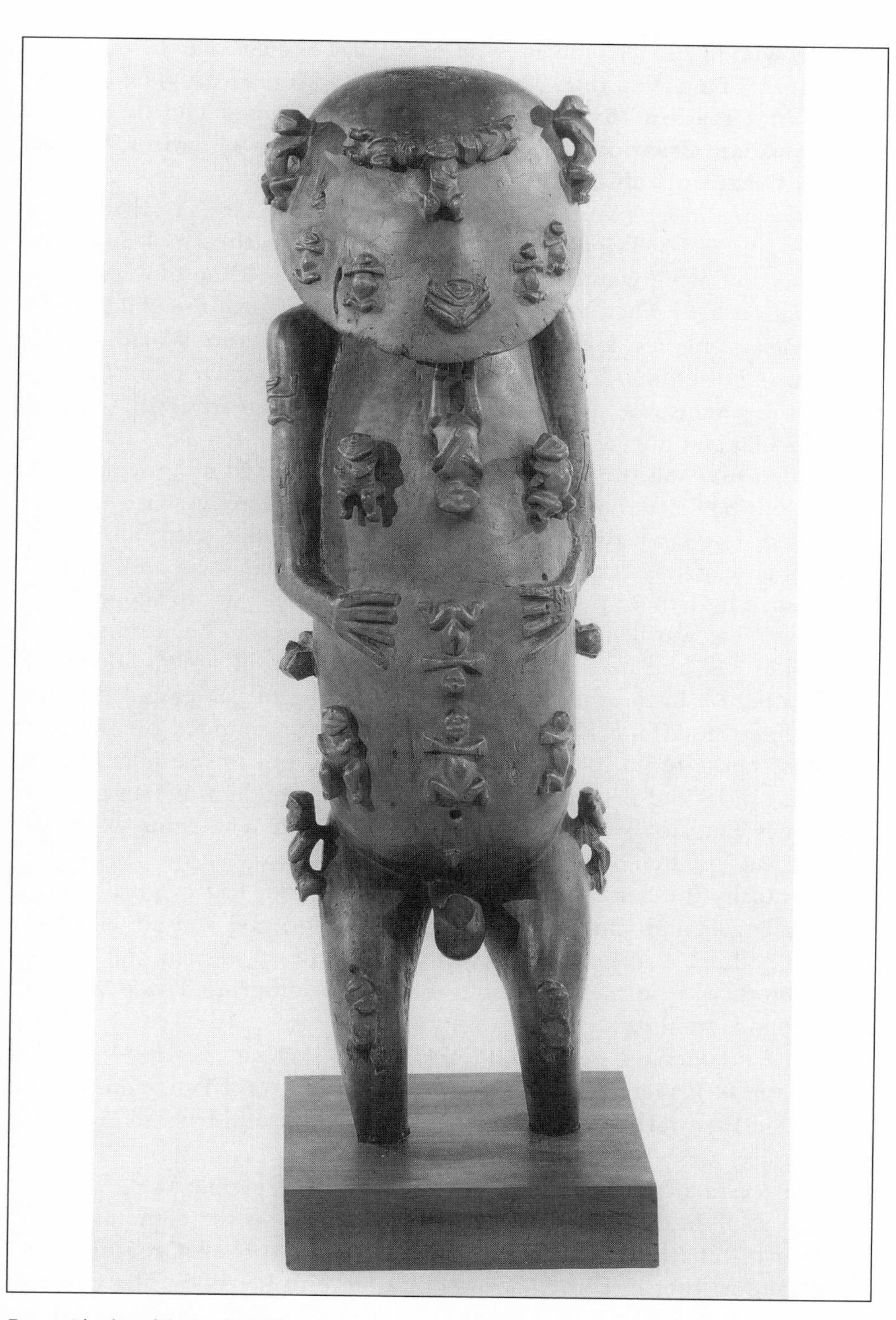

Rurutu islanders of the South Pacific carved an image of their creator god Tangaroa with beings emerging from his body to represent the act of creation.

instance, who first made woman—out of red clay. She became the mother of humanity by Tane, who then committed incest with their daughter (see also **Incest in Creation Myths**). The first male was named Tiki or Ki'i (see also **Hawaiian Creation; Maori Creation; Samoan Creation; Society Islands Creation; Tahitian Creation**).

There are more complex myths. Many Polynesians say that in the beginning there was Rangi and Papa (Heaven and Earth) coupled in darkness. Their offspring wondered about creation but could not see it because there was no light. They held council, and the angriest of the children, Tu-matauenga, suggested killing the World Parents (see also **World Parent Creation**). His brother, Tane-mahutu, suggested the simpler process of separating the parents (see also **Separation of Heaven and Earth**). Only the god of winds and storms voted against this plan.

Rongo-ma-tane, the god of cultivation, tried to stand up to separate the parents, but he failed. Tangaroa, ocean god of fish and snakes, tried next but also failed. The same thing happened when other gods tried, until Tane-mahutu, god of forests and flying things, stood on his head and, with his feet, pushed his father up from his mother as they cried out in agony.

Now there was light and the people were revealed. The wind god was angry, however, and he sent forth a terrifying series of storms and hurricanes to punish his brothers. The various offspring of Rangi and Papa, led by the forest god, Tane, and the ocean god, Tangaroa, argued over what to do. Some chose to go to land, some chose going out to sea. Tane gave his children canoes and nets so they might catch the children of Tangaroa, and sometimes Tangaroa swallowed up Tane's people. It is still this way today between the children of Tane and Tangaroa.

Eventually the fierce brother, Tu-matauenga, who had originally advocated killing Rangi and Papa, overcame the wind god and ate all of the original gods. Tu-matauenga is, in reality, the form called man, and after he ate his brothers, he turned them into food for his offspring. This is what the original gods are today.

Some Polynesians—especially in Hawaii and New Zealand—say that the last son of Rangi and Papa (Heaven and Earth) was Maui, who, like the Greek god **Prometheus** brought fire to humankind and tried to give them immortality.

Still another South Pacific myth says that the people live in a huge coconut shell. In the depths of the shell's interior lives the demon-woman, Vari-ma-te-tekere or Very Beginning (see also **Mother-Creatrix**). She made the first human being, Vatea, from a piece of her body. The father of humans and gods alike, Vatea was half man and half fish. His name means noon, and his eyes are said to be the sun and moon. Vatea was given the perfect middle land between brightness and gloom. Later he would marry the well-known Polynesian goddess, Papa (Foundation).

Out of another bit of herself, the Great Mother made Tinirau (Innumerable), who was also half man and half fish and was given the Sacred Isle (Motu-Tapu) as his home.

Out of her body the mother made many other beings and places for them to live. She even made Raka (Trouble), who controls the winds. Her last child was the beloved daughter, Tu-metua or Tu, to whom the moon is sacred. Her nephew was the famous Polynesian god Tangaroa (Taaroa).
Sources: Colum, 254–260; Freund, 126; Long, 58–63, 91–98.

Pomo Creation

In their **creation from chaos** and from nothing myth, the Pomo Indians of California say the creator was Old Man Madumda (see also **Creation from Nothing**). One day Madumda decided to make the world. He wanted advice from his older brother, Kuksu, so he plucked hairs from his head and asked them to lead him to where Kuksu was. He held up the hairs to each of the four directions, and it was to the south they flew. Madumda followed them on his cloud and spent the time smoking his pipe until he got to Kuksu's house. There, as is proper, the brothers smoked the pipe four times before speaking.

Then Madumda scraped skin from his armpit, rolled it up, and gave it to Kuksu, who placed it between his toes. Kuksu took some skin from his armpit, rolled it up, and gave it to Madumda, who placed it between his toes. Each blew four times on his little ball of skin, and then the two gods mixed the two balls with a little of their hair.

They stood up and faced the four directions and the up and down, then proclaimed the creation to come. Madumda took the ball and left as Kuksu sang the ancient creation song for the first time. Madumda sang, too, as he flew home on his cloud, with the ball strung through his earlobe. Then he slept for eight days, during which time the ball grew and became the earth. Madumda awoke and threw it into the air. Then he smoked his pipe and threw it, burning, into the sky, where it became the sun. In the new light he walked about creating things as we know them—the mountains, trees, valleys, rocks, lakes, seas, plants, animals, and so forth. By rolling the earth one way and then another, he made night and day.

One day he decided to make the people. First he made some stubby little people out of rocks, then some beautiful long-haired people out of his hair. He made some bird people out of feathers and some hairy deer people out of his armpit hair. In fact, he made all kinds of people out of all kinds of material. Finally he made naked people like us out of little pieces of sinew planted between some hills. Madumda gave these people their land and taught them how to eat and live.

In time the people began to misbehave (see also **Fall of Humankind**), killing each other and not caring well enough for their children, so Madumda sent a great flood to get rid of them (see also **The Flood and Flood Hero**). When they were all gone Madumda wished for a village (see also **Creation by Thought**) and there was one. He filled this village with people from his thoughts, but these people also went bad, so Madumda sent a great fire to destroy them.

Then he made new people out of willow wands. He taught them how to hunt with bows and arrows, how to make baskets, and how to eat before he went away to his northern home. These people also went bad and had to be destroyed—this time by ice. Then he made another willow wand batch of people and went away again.

After a time he was amazed to learn in his dream that these people, too, were all wrong (see also **Imperfect Creation**). Kuksu advised Madumda to destroy these people with wind, which he did. Only the ground squirrel escaped in his hole.

Madumda made new people out of willow wands. This time he made many groups that spoke different languages. He taught them their dances and ceremonies and he taught them how to behave and how to eat, weave, hunt, and grow things. He ordered the coyotes to watch over the villages. He also gave all of the animals their particular places. Finally he left the world, warning the people to behave properly. This is our last chance.

Source: Leach, 37–46.

Popol Vuh

The Popol Vuh—Book of the People—contains the sacred history of the Quiché Maya (see also **Mayan Creation**) from the beginning of time. The original version, in Mayan, was destroyed by the conquering Spaniards in the sixteenth century. The book was recomposed by a Mayan convert to Catholicism, who transposed the Mayan into Latin script. A copy of this manuscript was made by a Spanish priest, Francisco Ximenez, in the late seventeenth century. At the same time it was translated into Spanish.

Primal Mound

In some creation myths, earth began from a primal mound rather than from a cosmic egg. These creations tend to stress the fertility of the Earth Mother, the mound signifying the sacred place of birth (see also **Egyptian Creation**). The mound is also related to the many earth-diver stories in which the world begins from a small mound of dirt brought up from the depths.

Primal Waters

The primal waters are common to many creation stories, such as those of the earth-diver type. Like the primal mound, they are, in a sense, of the original Mother Earth—the maternal waters out of which the new creation is born.

Prometheus

There is a Greek story that has the Titan god, Prometheus (a potter), creating the first man and woman out of clay (see also **Creation from Clay**). It is said that Athena gave these figures life. Prometheus was also said to have given fire to humankind and generally to have championed them against Zeus.
Source: Graves (Vol. 1), 143–144.

Prose Edda

The Prose Edda or Younger Edda, as opposed to the **Poetic Edda** or Elder Edda, is also sometimes called the Snorra Edda after its author, Snorri Sturluson. With the Poetic Edda, it is the primary source for early Scandinavian (Norse)/Icelandic mythology (see also **Icelandic Creation; Sturlusun, Snorri**).

Pueblo Creation

There are many versions of the Pueblo emergence (see also **Acoma Creation; Creation by Emergence; Hopi Creation; Keres**

Creation; Laguna Creation; Zuni Creation). The Pueblo people comprise many groups of Native Americans who live in villages (pueblos) along the Rio Grande River in New Mexico. Most of these people speak the Keres language, but some speak Tewa (also Tiwa or Towa). The Zunis, Hopis, Acoma, and Laguna Indians farther west are also considered Pueblo peoples. All of these groups know at least one version of the following myth.

"Long ago when the earth was soft," all beings could communicate with each other, and the kachina (spirits) came in person to dance and sometimes even to fight. Coyote himself behaved in that golden age. The people say that in Old Oraibi it was possible to grow corn in a day, and that at Isleta water came out of rocks in profusion. The Keres mothers had seeds on their skins, seeds that grew immediately when planted. The people were perfect in those days and so was life.

It was Sun who called the people up from one world to the next. Eventually there were troubles, and now people are not perfect and corn does not grow so easily.

Source: Long, 57–58.

Quetzalcoatl

A primary god in the **Aztec creation,** Quetzalcoatl is the Feathered Serpent, the Phoenix-like resurrection god who has strong solar characteristics and is born of a virgin.

Quiché-Mayan Creation

See Mayan Creation; Popol Vuh.

Ra

Ra, also Re or Amen Ra, is one version of the Egyptian high god and creator (see also **Atum; Egyptian Creation; Khepri**).

Raven

Like Coyote, Raven is a creator-trickster figure. The many stories about him, especially among the Native Americans of the Northwest, form what can justifiably be called the Raven cycle. As the cycle moves east from the coast, Raven becomes less a creator and more of a villain. In the far north, among the Eskimos, he is more of a culture hero (see also **Chuckchee Creation; Coyote; Eskimo Creation; Trickster; Tsimshian Creation**).

Rhodesian Creation

See Wakaranga Creation.

Rig Veda

Dating from at least the second millenium B.C.E., the Rig Veda contains the earliest known versions of the Indian (Hindu) creation myths and is the most sacred book of Hinduism (see also **Indian Creation**).

Roman Creation

The Roman creation is markedly influenced by the Greek myths of Hesiod, and it has the familiar elements of the separation of earth and heaven, the war in heaven, and the Fall. It is somehow more orderly and less supernatural than the Greek creation, however. It reflects a more scientific and skeptical age. The storyteller here is Ovid (43 B.C.E.–18 C.E.), the great Roman poet whose *Metamorphoses* is the primary source for Roman mythology. Ovid's myth is a creation *ex nihilo* or from chaos (see also **Creation from Nothing**).

In the beginning was the formless and random mass we call chaos. There was no sun, there were no stars, there was nothing of permanent shape, and everything got in the way of everything else.

At some point, a profound natural force, a god, separated earth from heaven (see also **Separation of Heaven and Earth**), and earth from the waters, and developed harmony (cosmos) out of chaos. Ether went up to the highest place; then came air and finally heavy earth, which attracted all the gross elements of creation, sank to the bottom, and was held in place by the surrounding seas. Earth itself was shaped into a great ball and arranged into the areas of the world.

When everything was ready, the Creator, or some say Prometheus, decided to make humankind to watch over creation. The first man was made out of rainwater and some of the higher heavenly elements still found on earth; he was made in the Creator's image.

At first there was the Golden Age, when everything was perfect in heaven and on earth. Humans ate freely of the abundant fruits of the earth.

Later there were struggles in heaven, and Jove (Jupiter) replaced Saturn as the king there. The seasons developed on earth, and a bit of hardship. Agriculture was practiced. This was the Silver Age.

Then came the Bronze Age, during which humans were good still but during which they turned to the arts of war.

The last age was the Iron Age, in which goodness was overwhelmed by evil. There was war, greed, hardship, and disloyalty. No one could be trusted, and love died among humans. There were also wars in heaven during this time, between the forces of Olympus and those of the ancient giants. After settling the affairs of heaven, the gods decided to destroy humanity on earth. This they did with a great flood (see also **The Flood and Flood Hero**). Only the guiltless Deucalion and Pyrra were saved in a boat so they could be the parents of a new human creation.

Source: Ovid, Book I.

Rumanian Creation

This is an *ex nihilo* creation myth of ancient Rumania (see also **Creation from Nothing**).

God made Heaven, and then, after measuring the space underneath with a ball of thread, he began to form the earth (see also **Creation by *Deus Faber***). A mole asked to help, and God gave him the thread to hold while he wove the patterns of the earth. Sometimes the mole would let out too much thread, and finally the earth grew too large for the space under heaven. The mole was so upset that he hid under the earth. God sent the bee to look for him; he wanted the mole's advice on what to do about the mistake (see also **Imperfect Creation**).

The bee found the mole and he just laughed at the idea of advising God. The bee, however, hid in a flower and overheard the mole mumbling to himself about what he would do if he were God. "I would squeeze the earth," he said. "That would make mountains and valleys and make it smaller at the same time."

When the bee heard this, he went directly to God and told him. God did what the mole had said, and everything fit fine.

Source: Leach, 31–33.

Sahaptin Creation

See Salishan/Sahaptin Creation.

Salinan Creation

In the creation of the Salinan Indians of California, there is a flood myth in which Eagle takes the place of Raven or Coyote as the animal possessed of supernatural creative power. The flood gives birth to an **earth-diver creation** (see also **The Flood and Flood Hero**).

Because Eagle was so grand and so powerful, Sea Woman was jealous of him, and she came toward him with her great basket that contained the seas. She poured the water onto the land until only the top of Santa Lucia Mountain was left. Eagle gathered the animals there and borrowed Puma's whiskers to make a lasso, and he lassoed the sea basket. Sea Woman died and Eagle sent Dove to get some new earth. Out of this mud Eagle made a new world. He also made a man and a woman out of elder wood, but the new creatures had to be taken to the sweat house by Prairie-Falcon and breathed on by Eagle to be given life.
Source: Sproul, 236.

Salishan/Sahaptin Creation

These Northwest Indians have a **creation from chaos** myth that contains the familiar motifs of the world tree (see also **Yggdrasil**) and creation from a lump of clay (see also **Creation from Clay**).

The Sky Chief made the earth out of a lump of clay, and he rolled it out like a piece of dough until it was the size it is now. He covered the earth with soil and made the heavens and the underworld as well. He connected them all with the world tree. He made animals and a man—a wolf-man. He then made a woman from the man's tail. These were the Indian ancestors.
Source: Sproul, 243–244.

Samoa Creation

The Samoan Islands are inhabited by Polynesian people whose myths contain elements and characters that are found in most Polynesian myths (see also **Hawaiian Creation; Maori Creation; Polynesian Creation; Society Islands Creation; Tahitian Creation**).

Some Samoans say that Tangaloa-Langi, the creator, lived in a cosmic egg (see also **Creation from a Cosmic Egg**). When the egg broke, the pieces of the shell fell into the waters and became the Samoan Islands.

In the best-known Samoan creation, however, the *ex nihilo* creator is Tangaroa, the ocean god for many Polynesians, but the supreme being for the Samoans (see also **Creation from Nothing**).

Tangaroa lived in space alone before there was any form to the universe. He stood still once, and a rock grew. He told the rock to split, and it did (see also **Creation by Word**). Many other rocks came along, too, representing various phrases and ideas still used by Samoans.

Tangaroa hit the original rock and it gave birth to the earth and sea. Then the various rocks spoke to each other. Tangaroa spoke to the main rock many times, bringing forth fresh water, the sky, space, height, and other things and ideas. He also called up maleness and femaleness, Man, Spirit, Heart, Will, and Thought. Then he left his creation floating about aimlessly after instructing the rock.

Tangaroa told the rock that spirit, thought, will, and heart were to come together in man, and they did that. Man was to join with Earth to make a couple—Fatu (male) and 'Ele-ele (female). They were to populate a certain part of the world.

A certain rock-phrase, "Chief-to-Prop-up-the-Sky," was told to hold up the sky over the earth, but he was not very successful until he made posts.

Now Tangaroa-the-Creator made Tangaroa-the-Immovable (chief of the sky) and Tangaroa-the-Messenger (his ambassador to the other heavens).

Night and Day lived in a lower heaven and produced Manu'a, Samoa, Sun, and Moon as offspring. The messenger called Night, Day, and their

children to a council with the Creator and the Immovable. There they were told that Manu'a and Samoa should go down and become chief over the descendants of Fatu and 'Ele-ele. Sun and Moon were to go down, too, to follow Night and Day.
Sources: Freund, 56–57; Sproul, 346–347.

Samoyed Creation

These Siberian people tell a creation story much like that of the Buriat and Altaic peoples (see also **Altaic Creation; Buriat Creation**). In this "How-the-Dog-Got-Its-Fur" myth, we are told that man and the dog were both created naked, and that the devil gave the dog hair by patting him.
Source: Leach, 201.

San Cristobal Creation

The Melanesian people of San Cristobal say that Agunua made the earth and the waters *ex nihilo,* that he made the storms, and that he created humans (see also **Creation from Nothing**). He gave his brother a yam and told him to plant it, and it produced the fruits the people like—banana, almonds, and coconut. The brother burnt up some yams once, however, and this made some fruit poisonous (see also **Banks Island Creation; Fiji Island Creation; Imperfect Creation; New Hebrides Creation; New Guinea Creation**).
Source: Leach, 176.

Seneca Creation

In the Seneca creation we find the familiar earth-diver form (see also **Earth-Diver Creation**) and the typical Iroquoian motif of Star Woman (see also **Cherokee Creation; Huron Creation; Iroquoian Creation**).

There was a time when water was everywhere and was populated by ducks, loons, other waterbirds, Turtle, and Toad. In those days, the people lived in the sky with the Great Chief. One day the chief's daughter fell sick and began to die. A wise man learned in a dream that she should be placed next to a tree and that the tree should then be dug up. He told the chief about his dream, and the chief followed the dream's instructions.

Then a man came along who resented the digging up of the tree and kicked the girl into the hole. Suddenly she was floating down through space. Seeing what was happening, the birds rose and formed a soft net with their wings and caught the girl. When they got tired, they put her on Turtle's back, but he got tired too. The birds realized the girl would need something to rest on, so Toad dove down to the bottom of the waters and brought back a bit of soil. She placed it on Turtle's back, where it began to grow. Turtle's back grew, too. Soon there was the earth for the girl to live on. She was happy there; she made a little house, and soon she produced a baby girl.

The woman and the girl worked the land. Soon the daughter had twins boys, Flint (Othagwenda) and Sapling (Juskaha). Star Woman did not like Othagwenda, so she put him in a tree. She taught Juskaha how to make things and hunt. Soon she noticed that he would come home without his bow and arrows. It seems he was giving them to the twin in the tree. Finally he brought his twin home with him. They stayed there together for a long time. Then they decided to enlarge the earth. Othagwenda made Mosquito and rough land. Mosquito was huge and could even chop down trees. Juskaha was horrified. "This is a terrible animal; he might kill the people we plan to create," he said, and he rubbed the animal down to his present size. As for Othagwenda, he did not like his brother's creations—big fat animals, rich syrup-dripping maples. "These animals must be made harder to catch," he said. He made the animals thinner and faster, and then made the maples drip sap that had to be boiled into syrup.

Finally, the two brothers fought. Juskaha killed Othagwenda, but it was too late for the good brother to change the bad brother's work.

Source: Leach, 82–87.

Separation of Heaven and Earth

This separation is a basic motif of creation myths. It assumes that until there is differentiation there can, in fact, be no creation as we know it. The world parents—the unified or coupled unity of earth and heaven or sky—are suffocating for the potential world caught within the potentially divisible unity (see also **Creation from Division of Primordial Unity; World Parent Creation**). In fact, the unified heaven and earth are analogous to chaos in other myths (see also **Egyptian Creation; Geb and Nut; Krachi Creation; Minyong Creation**).

The Serpent in Creation

The serpent is a traditional working companion of the Great Goddess in the most ancient myths. It is sometimes the goddess herself and sometimes the phallic symbol of earth's fertile powers. In later patriarchal creation myths, the serpent becomes something to be conquered, something associated with the unformed mysteries of the dark goddess—Earth—as opposed to the rational and light-bringing sky god (see also **Babylonian Creation; Hebrew Creation**).

Shamanism and Creation

The shaman or medicine person is the man or woman of a given tribe who has learned the ways of the spirit world, who can dive down into that world—literally or figuratively—to find the source of curing, or new creation. There is an obvious connection between shamanic practices and creation myths, especially of the earth-diver type, where the animal dives down to find the source of creation. In fact, shamans often wear animal masks (see also **Earth-Diver Creation**). Shamans also resemble tricksters in that they are transformers of reality with magical creative powers.

Shilluk Creation

The **world parent creation** myth of the Shilluk people of the Sudan features Jo-Uk (the creator) and the Sacred White Cow. Together they produce man.

When the world was new the great creator Jo-Uk made the sacred white cow, who gave birth to a son she named Kola, who in his turn produced a son named Ukwa. Ukwa's two dark virgin wives came out of the Nile, the holy river, the river from which the sacred white cow originally emerged. Nyakang, one son of Ukwa, was a tall, blue-black warrior. He travelled south and founded the Shilluk nation, which stood on the marshes of the Upper Nile. Nyakang became the first ruler and demigod of the Shilluk.

Source: Freund, 5–6.

Shoshonean (Luiseno) Creation

The Shoshonean Indians, often referred to as the Luiseno people after the Mission of San Luis Rey near Los Angeles, tell a complex creation story that contains many familiar motifs, beginning with creation from chaos (see also **Creation from Nothing**). It is a story that has similarities with the Hawaiian myth of the gradual coming of light (see also **Hawaiian Creation**).

In the beginning there was only Kevish-Atakvish (spacevoid) or Omai-Yamai (nothingness). Then things began to fall into forms. Time came and the Milky Way. There was no light yet, but there was a creative stirring.

Kevish-Atakvish made a man, Tukmit, who was the sky, and a woman, Tomaiyovit, who was the earth. They could not see each other, but brother and sister knew each other and they conceived and gave birth to the first elements of creation (see also **Incest in Creation Myths**). They produced the valleys, mountains, stones, streams, and all things that would be necessary for worship, ceremonies, and cooking. From the earth came Takwish, the terrifying meteor, and his son, Towish, who is the immortal soul of humans. Wiyot also came forth, and from Wiyot came the people. It was still dark.

The Earth Mother made a sun, but it was too bright and had to be hidden away since it frightened the people.

The people made more people and they followed the growing earth as it stretched southward. They came to Temecula, where the Earth Mother brought out the sun again. The people raised it up to the sky, where it followed a regular path and was not so frightening.

At Temecula, it is said by some, the father of the people, Wiyot, died. Because Frog hated him for the legs he had made for her, she spit poison into his water. After drinking the poison, Wiyot announced that he would die in the spring. Before he left, he taught the people what they needed to know. When he died, a great oak tree grew from his ashes (see also **Yggdrasil**).

Now Wiyot visits the people each night; he has become the moon and is the center of their celebrations: "Wiyot rises," the people cry as they dance for him.

Sources: Leach, 60–63; Weigle, 202–205.

Siberian-Tartar Creation

This creation is really a version of the Altaic myth (see also **Altaic Creation; Buriat Creation; Samoyed Creation**). In this version the great high god calls Pajana, the creator god, to heaven to give him the magic of life. While he is gone, Pajana leaves the furless dog to watch over the lifeless objects he has created, but the evil Erlik spits on them and they have to be turned inside out before life can be given to them.

Source: Leach, 201.

Sioux Creation

Among the various branches of the Sioux tribes of the American Plains are several types of creation myths. The Brule Sioux of South Dakota tell a story that begins with a flood, suggesting a forgotten creation of earlier times (see also **The Flood and Flood Hero**).

The first people were attacked by the great Water Monster, who sent a flood to kill them. The people tried to escape by climbing the steep hill in the middle of the Brule land, but the water found them and drowned them. All that was left was a pool of blood, which became the sacred red pipestone quarry. The pipe later made from that rock—the blood and bones of the first ancestors—is sacred, too. When it is used it has great power; the breath of the old people is in the smoke that comes from it.

After the flood the Water Monster was turned to stone like the people. She became the terrifying place called the Badlands.

There was one person who managed to escape the flood. She was a fine young girl; she was rescued by Wanblee, an eagle, and taken to the highest spot in the Black Hills, the tall tree that was the eagle's home. There the girl became Wanblee's wife and gave birth to twins, a boy and a girl, who later became the parents of the Sioux nation. The eagle was, of course, the messenger of the Great Mystery, and the Sioux are proud to be called eagle people.

The Lakota Sioux say that in the beginning First Man emerged like a plant from the Great Plains earth (see also **Creation by Emergence**). At first only his head was visible, and he looked around at the nothingness that surrounded him. There were not yet any rivers, mountains, grass, or animals. Gradually First Man pull himself up out of the soil until he stood on the soft earth. It was the sun that gave solidity to the earth and strength to the man.Out of this man came the Lakota people.

This is a recent creation myth told by a Brule Sioux medicine man, Leonard Crow Dog, on the Rosebud Reservation in 1981. It came to him, he said, in a dream during a vision quest, a ritual spiritual search.

> This story has never been told. It is in no book or computer. It came to me in a dream during a vision quest. It is a story as old as the beginning of life, but it has new understandings according to what I saw in my vision, added to what the grandfathers told me—things remembered, things forgotten, and things re-remembered. It comes out of the World of the Minds.
>
> Some people say we are descended from Adam and Eve, but there was no Adam or Eve in our creation. Some people try to tell us that we were born with the burden of original

sin, but that is an alien white man's concept. Sin was not in the mind of the universe of our creators or the created.

When this world came into being seven million eons ago, it was composed of numberless hoops, skeletons with no substance. Land, the whole earth, had yet to be made. All was orbits within orbits within orbits. The world on which we are sitting now, our earth, was made up of sixteen sacred hoops. There was no earth, no land, but there were planets and stars. Above all there was the great sun. He controlled all the orbit powers. He had the sole power to communicate, to talk planet-talk between the universes, stars, and orbits.

The sun had seven shadows, and in them he recreated himself. The seventh shadow was the important one. The sun looked at it and saw its design was different. The shadow was the creator of the red man's land.

Then the great sun called to all the orbits, planets, and stars: "Come to the sixteen hoops! Come to the sixteen hoops!" and they all went to the place the sun had appointed and made earth-plan talk. The sun would not allow them to leave until they were done. And that great ball called earth, the earth planet himself, said to the sun: "Instruct us in the way of the universe." For this purpose and for this reason the orbs and the orbits talked to each other. They related to each other and that was the first relation-making feast, the first *alonwanpi* of the universe.

And one of the orbits, the east, asked the sun: "Why have you called us to come here? What have you called me for?"

"I have called you because you shall take part in this creation. You will breathe into these sixteen hoops. You will breathe into them with your Takuskanskan, the moving power, the quickening power which is part of the Wakan Wichohan, the big sacred work we must do."

And the orbit which is called the south asked the sun: "Why did you call me? I have come with my planets and my orbits and you must tell me what to do." In this manner the orbits talked.

Then the west asked: "Sun, why have you called all the orbits and planets here? What is the purpose?"

"I have called you here in a sacred manner, for a sacred purpose: To help me make this earth, this land. To breathe into these sixteen hoops."

The north said: "What is that thing you called land, what is this thing you call earth? What have I to do with it?"

The great sun replied: "You are the living moisture; you are the atmosphere; you are the north. You will be the caretaker of this earth. You will make the seasons in all eternity."

And the power from the north answered: "Hou, hou!" This was the echo of echos of all the universe, and it reverberated throughout the hoops and orbits.

So the sacred four-direction powers breathed their life-giving breath into this earth we are sitting on. The sixteen hoops were still skeleton hoops; you could see through them, walk through, float through; they had no substance yet.

The sun again called all the powers and planets to crowd around the earth and breathe into it, and this was the beginning of the red man's life. All the powers of the universe participated in its creation, but there arrived among them in a whirlwind an unknown power right out of the center of the universes. Its name was Unknowingly, and it also breathed into the sixteen hoops. All the powers breathed fire and the other elements into this land, and when they had finished, one and a half million eons of creation had passed.

The sun looked at the earth. Everywhere he saw beauty and light. He saw the art designs of the universe, the creation art of the planets, and land painting. The sun gathered parts of all the riches of all the universes and put them into this newly created world; nothing was wasted.

But the earth was bare. It was a bald-headed world. No life was on it yet; it was rock, a far-shining crystal.

The Great Unknown Power, the Grandfather Power, Unknowingly, was part of the sun and the sun was part of him. Unknowingly was seen-unseen and had many forms. He spoke: "Ho! Aho! Now it is done. This is the Great Way of the Great Spirit talking." And of the earth he said: "This will be my seat. This will be my backrest." In the earth he planted the seed of life, a planting that took half a million eons of creation time.

First Unknowingly planted trees, the kind that never change, that are always green: the pine and the cedar. They are the green relations of the universe, and we still use the cedar as incense in our ceremonies. In his mind this tree planting was done in the blink of an eye, but it lasted a million and a half eons of creation time. At that point the sun did not move yet, did not rise and did not go down, just stood in one place. The sun looked at the earth covered with green and said: "It is beautiful. I am satisfied." Then the great sun made the four seasons for north to take care of, and when he had finished, another half-million eons of creation time had passed. And no birds had yet been created; just our green relations.

The trees spoke to each other. Every day and every moment they were talking, and they are still talking now in an unknown language which humans do not understand. When a little child emerges from the womb, the first thing it does is to cry and cry. It too is speaking in an unknown language—tree language, universe language, survival language. Though the newborn later forgets, he knows at birth that we have to survive to take care of this world, to live in a sacred manner after the original instruction.

When three million eons of creation time had passed, the great sun looked down from his orbit and thought: "This is unique. Everything moves in the Great Way. Caretakers, the sacred four directions, have been appointed, and they are doing well what they are supposed to do." And he looked at a tree and saw that a big branch was broken off. He said: "Ton, Ton, Tonpi. Birth giving. It's time for creating people, for forming them up in pairs."

Don't call us Indians: call us Birth People, because that is what we are.

The sun thought: "Everything looks nice, and birthing is about to take place, but somebody should be the caretaker of this Birth people land. The four-direction powers already take care of the planet, but I want a special caretaker for the hemisphere upon which I shall put the red man." At that time he did not think of it as Mother Earth but as the Planet of the Universes, the Orb of Planification. Because there was no mother yet, no man or woman; just the colors of the four directions and the plants, and the intelligence of powers, the intelligence of Tunkashila.

The great sun called loudly: "Unknowing, you always arrive unknowingly. Come unknow[l]ingly from your seat." And Unknowingly arrived with lightning and with powers that no human could scientifically analyze, that could not be computed, powers sacred and secret, the oldest, the most innate. Unknown was a shadow who spoke with lightning, with thundering. The great sun, *anpetu-wi,* still stood idle, fixed in his place from the moment of creation. Then suddenly, at billions of miles an hour, the sun began to move. Moving, he released glowing gases, the energy of the fire without end, life-giving warmth. Unknowingly was right beside him at that moment of creation time. (Were he and the sun one? Were they two? Was he the sun's seventh shadow?)

Unknowingly said: "Now we are going to make a human out of all these elements. We will take the vein of the cedar tree to create a man who will be the caretaker of this land. His name shall be Ikche Wichasha—the Wild Natural Two-Legged, the wild, free human. Unknowingly was the seventh shadow of the sun, and he spoke the lightning language to communicate his wishes. If the shadow walked through this room here, you couldn't see him, but you would somehow feel his presence and you would have a new vision.

Unknowingly called the whirlwind. "Yumni-Omni, Tate Yumni, arrive!"

Whirlwind arrived with a thundering moan—the earth-birthing sound. The sun, from his eye of eyes, his eye of the universe, made tears flow. When one tear hit the earth, it turned into a blood clot, a *we-ota.* It was as yet only a shadow, but for four generations this shadow developed itself. The whirlwind enfolded him, hit him, helping him to become a body. He was We-Ota-Wichasha, Blood Clot Boy, and he was almost seven feet tall. When the whirlwind hit him, supernatural knowledge went into him, as well as the power of speech and the knowledge of language. And when Blood Clot received these powers, he became a man. The sun was content, saying: "Now a caretaker has been created for this land."

We-Ota-Wichasha developed not only into one man, but into seven nations of the seven ore colors. Today we have only four colors—the red man, the white man, the black man, and the yellow man. What happened to the other three kinds of men? Where did they go?

One was Kosankiya—a great planet, with plants, with animals, with humans. Kosankiya is the darkness of every blue. He said: "I shall be the nest maker. I shall be the upholder of the dome. I shall be the blue sky." He is still here, whether it is day or night. That vault above us makes himself dark at night, blue during the day.

And where is the second one? His name is Edam, Hota Edam, Hotanka—the Great Voice blazing forth. Where does he come from? He is floating in the voids. He is red, an art design. You can see him among the thunder clouds sometimes. And he is the Wakinyan, the great thunderbird, the winged part of the sixteen sacreds. He is still here.

And still one is missing; where is he? Look carefully, for he is the spirit of the land, the yellow spearhead of the earth powers. He is Wo-Wakan, the supernatural.

Together with the four races of mankind, the Above, the Below, and the Winged Spirit form the seven generations. None are missing. And we are part of them. They include us, they include everything; even a pebble or a tiny insect is gathered up in the sacred hoop.

Now, the sun had given Blood Clot Man the intelligence of the divine human being. He was a medicine, for the sun had shed tears and sweated as during a sweat-lodge purification. Out of the winds, out of the whirlwind, out of the sacred breath of the universe Blood Clot had been made. He was not created in nine months, like the child you and your woman begot, but over millions of years. Yet even in your baby, a little of that lightning power and star breath is being passed on.

At this time the earth was a crystal inhabited by a great intelligence and overblanketed by the sun and the shadows he had created. Its shining center was crystal, glass, and mica, but it was solid now. You could not pierce it or walk through it, for the skeleton had been covered with flesh, green flesh. Next Wakan Tanka, Tunkashila, formed animals in pairs, to give their flesh so that man could live. And then it was time to create woman. There was no moon then; it was still the period of sacred newness. The sun again called all the planets and supernaturals, and when they had assembled, the sun, in a bright flash, took out one of his eyes. He threw it on the wind of his vision into a certain place, and it became the moon. And on this new orb, that eye-planet, he created woman. "You are a planet virgin, a moon maiden," he told her. "I have touched you and made you out of my shadow. I want you to walk on the earth." This happened in darkness at the time of a new moon.

"How will I walk over to that land?" asked the woman. So the sun created woman power and woman understanding. He used the lightning to make a bridge from the moon to the earth, and the woman walked on the lightning. Her crossing took a long time.

Now the maker of the universe had created man and woman and given them a power and a way that has never been changed. Doing that, the sun had used up another million eons of creation time. He instructed the woman in her tasks, which she accomplished through her dreams, through her visions, through her special powers.

The Great Spirit had created the woman to be with the man, with We-Ota-Wichasha—but not right away. They had to make contact slowly, get used to each other, understand each other for the survival of their caretaking. Tunkashila let blood roll into her. She walked on the lightning, but she also walked on a blood vein reaching from the moon to the earth. The vein was a cord, a birth cord that went into her body, and through it she is forever connected with the moon. And nine months of creation were given to her. At first she was without feeling, for love was created in her and inside the man long after their bodies had been formed. They did not live as we do today but were a part of the land, taking care of it even while it took care of them.

The man and the woman began to communicate with each other and talked for many years. Then inside them a feeling emerged. Even before they touched each other they felt a vibration, womb understanding. So by the powers of the great sun, by the powers of Tunkashila, it was given to them to understand that they were man and woman, creators themselves. That understanding came to the man through lightning, through the sun blood that was in him, and it came to the woman through that birth cord which connects her to the moon and whose power she still feels at her moon time.

"You are the caretaker of the generations, you are the birth giver," the sun told the woman. "You will be the carrier of this universe."

The man and the woman did what they were meant to do after sacred nature's way, and twins were born to them, two little boys. They were not born in a hospital or a tipi but in the natural way, the woman crouching, gripping her birthing stick, with a soft deerskin waiting to receive her womb offerings. The sun dome was their dwelling, not a tipi or a house. Their roof was the sky vault. And that is why we, the red people, the Ikche Wichasa, are the oldest people on this hemisphere, living here since the beginning of time with the understanding and power given to us.

And the moment they were born, the twins already had that understanding and power. When they were old enough (and they grew faster than humans do now), they climbed to the top of a high hill for their vision quest, which lasted sixteen days and sixteen nights. They did not purify themselves in a sweat lodge before and after they cried for a dream, because everything was still pure. On the mountain one of the twins heard the voice of Tunkashila and answered, "Hou!" And Tunkashila showed him the path to making and keeping a flame—*peta owihankeshni,* the fire without end. And ultimately from the vision came the first sweat lodge. The one twin received this great vision, the other twin a lesser one. And each of them followed his own dream.

We-Ota-Wichasha and First Woman begat other children, boys and girls from whom sprang many nations. The twin who had received the great vision also had a son, begotten with the help of a sunbeam. When that son was old enough he too went up on a high hill for his vision quest, and the hill turned itself into a nest. "Ikcheha, Ikchewi, Ikche Wichasha, that will be your name," he heard a voice saying, "Ikche Wichasha, that is who you are,"—and that is what we Sioux have called ourselves ever since. Mark what is in this name: *che,* the male organ; *wi,* the sun; and *sha,* red. Together they mean "wild, common man," a natural free human, an earth man. But all those syllables and meanings are put in to show that we are the original red sun people. Ho He!

And the son of the twin made the first fire and built the first sweat lodge. Then he went to his parents and said, "I must leave you. I am appointed to take care of the winds of this universe." He began walking up the hill on which he had performed his vision quest and, before the eyes of his father and mother and of We-Ota-Wichasha and First Woman, he turned himself into an eagle. The eagle-son flew off with a gift from the Great Spirit—the four seasons. They accompanied him in the shapes of the bald eagle, the spotted eagle, the golden eagle, and the northern eagle.

We-Ota-Wichasha and First Woman saw their grandson fly away, circling higher and higher. And they went up to the mountaintop which had become a nest and found that he had left them gifts. A bow and arrow were lying there, and a rock, a spider web, and a gourd rattle. They found a fire stick and a small fire burning brightly. The eagle had scratched it out of the rock with his claw, striking a spark from the flint.

These things had been shadows out of a vision, and eagle-son's understanding had brought them into being, making them real. All the sacred survival things fitted themselves into the hands of We-Ota-Wichasha and First Woman, and through them were given the red man, together with the knowledge of how to use them. And when these First Parents brought the things back to their small camp, they found it swarming with people in many camp circles of many tribes. And to them all, We-Ota-Wichasha and First Woman imparted the vision and the dream and the sacred things and the understanding. And at that moment the seven million eons of creation were ended.

Sources: Erdoes, 93–95; Weigle, 138.

Skagit Creation

This myth is the **creation from chaos** story of the Skagit Indians of the Pacific Northwest.

In the beginning when the world was completely flooded, Suelick made four powerful brothers named Schodelick, Swadick, Hode, and

Stoodke. Suelick told his brothers to create earth and people, so they left Suelick and their home. Schodelick, the eldest, who stands for something round or perhaps even a canoe anchor, came to Skagit country and made man, woman, and some land. He also created fish for the rivers and lakes there and showed the man and woman how to fish. He taught them how to clean and eat the fish also. Then Schodelick created all things that live: the trees, plants, and all the other animals. Schodelick also taught man and woman how to use these. After finishing creation Schodelick went to the waterfall near Marble Mount and told his brothers about his work. He told them he would be in the water near the big rock under the falls, and he dove into the water and sang and swam for a long time. He still lives there and represents the greatest power for the Skagit Indians. You can even walk by this place where he lives, and if you are there early enough in the morning, you might hear his song.

When Schodelick first dove into the water after creation he was hungry. Thus the Skagit people, who live by the power search or vision quest, also dive into the water hungry after many days of fasting in order to hear Schodelick. Without the power received from this search, the Indian has no purpose and does not live long. All the young Skagit people search for this power.

Schodelick's three brothers went to Okanagan, where they created the earth and people there, as well as the trees, plants, and animals (see also **Okanagan Creation**). They taught these people the use of all the things in creation. Hode, one of the brothers, represents fire to some, and the people he helped create worship fire. They become wild and can return to this world only by beating their heads in the fire.

This is the true story of the beginning of these people.

Source: Rothenberg, 89–91.

Sky Father

The Sky Father is the complementary figure to the Earth Mother in many creation myths. Often he achieves a state of union with the mother and must be separated from her so creation can continue (see also **Creation from Division of Primordial Unity; Separation of Heaven and Earth; World Parent Creation**). In some myths he is the lone creator, the Old Man of so many Native American myths and God, the creator of monotheistic religions.

Snohomish Creation

This is the *ex nihilo* creation myth of the Snohomish Indians of the Pacific Northwest (see also **Creation from Nothing**). It contains hints of the **world parent creation** myth as well.

The Creator did his work from the east to the west, giving a different language to each group of the people he made. When he got to Puget Sound he liked it so much he decided to stop there. This is why there are so many different languages in Puget Sound and nearby.

The people complained about the low sky on which they constantly bumped their heads. The elders of the tribes met in council and decided to push the sky up. At an appointed time, someone would shout, "Ya-hoh," meaning lift together, in all of the Puget Sound languages, and the people would all press upwards with huge fur trees. It was difficult, but the people did use this method to push up the sky to where it is now.

There were three hunters, however, who knew nothing about the skyraising. They were chasing four elks over where the earth meets the sky, and they followed the elks into the sky when they leaped there. When the sky went up, so did the elks and hunters. They are still there now. The hunters are the handle of the Big Dipper, while the four elks are the bowl. A little star nearby is the dog of one of the hunters.

Others were caught in the sky that day—a little fish and six men in two canoes—and they are stars now too.

The Puget Sound people still say "Ya-hoh" when they want to lift something together.

Source: Erdoes, 95–97.

Society Islands Creation

The familiar Polynesian god, Ta'aroa—also Tangararoa, Tangaroa, or Tahitumu (Original One)—is the creator in the Society Islands creation story. He made everything, beginning with himself (see also **Creation from Nothing**); and he was born of a shell, which he turned into the world. The shell can also be seen as an egg (see also **Creation from a Cosmic Egg**).

In the beginning there was only darkness in the shell until Ta'aroa created himself out of his aloneness and broke out. Then he turned the shell into the sky and floated about in space for a time before entering a second shell. After a while he broke out again and made that shell into the earth. Then he went back to Ramia, the sky and first shell, where he still lives (see also **Tahitian Creation**).

Source: Eliade (A), 87–88.

Sophia

Sophia is the Greek name for **Wisdom,** the female companion of Yahweh at the creation as described in the Book of Proverbs. In the Gnostic tradition, Sophia was God's mother, the Great Virgin Mother in whom God was concealed before the beginning. For some Gnostics, Sophia was said to have been born of the female essence, Sige (Silence), and to have given birth herself to the male Christ and the female Achamoth. Achamoth produced Ildabaoth (Jehovah). When Ildabaoth denied humans access to the fruit of knowledge, Achamoth sent her own spirit as Ophis, the serpent, to teach humans to disobey Jehovah. The serpent was also seen as Christ. Later Sophia sent Christ to enter the man Jesus when he was baptized, and still later, Sophia and Jesus married in heaven (see also **Gnostic Creation**).

Sophia was suppressed in Western Christianity, but in Eastern Orthodoxy she was greatly revered. Her shrine was in Constantinople, a church built in the sixth century C.E. that Western Christians "renamed" for a martyr of the same name, St. Sophia.
Source: Walker, 951–952.

Spider Woman

Spider Woman is a popular creatrix among the Hopi and other Southwestern tribes (see also **Hard Beings Woman; Hopi Creation; Navajo Creation; Thinking Woman**). She is either the assistant to the supreme creative power or the personification of that power (see also **Spiders in Creation**). She is sometimes Spider Grandmother, as in the Hopi village of Old Oraibi, and she has a human counterpart in the old women of the various Pueblo clans, who weave the stories of the creative past for the children even as the Spider Fathers and Spider Grandfathers weave the ceremonial shawls in their underground kivas.

Spiders in Creation

Spiders, as spinners (spinsters) or weavers—that is, natural creators—are popular in creation myths, especially those of the American Southwest (see also **Hopi Creation; Krachi Creation; Lakota Sioux Creation; Navajo Creation; Papago Creation; Pueblo Creation; Zia Creation**). The spider can be male in the myth (see also **Inktomi**) but is usually female (see also **Spider Woman**). The spider figure often has a **trickster** aspect (see also **Inktomi**). The basis for the metaphor is the spider's ability to weave a clever, beautiful, intricate, and dangerous web.

Spinsters (Spinners) in Creation

See Spiders in Creation.

Sproul, Barbara

Barbara Sproul is the compiler of a significant collection of creation myths, *Primal Myths: Creation Myths around the World* (1979).

Struggle between the Gods

The struggle in heaven between the gods or between so-called high gods and more primitive first gods or giants is a frequently used motif in creation myths. As creation on earth is frequently imperfect (see also **Imperfect Creation**), the same can be said of heaven, which reflects earth in many ways. Sometimes the war in heaven represents a change in theology among humans, a change, for example, from an earth-oriented goddess religion to a heaven-oriented god religion (see also **Babylonian Creation; Greek Creation**).

Sturluson, Snorri (Sturlason)

Snorri Sturluson was the compiler of the **Prose Edda** (see also **Icelandic Creation**).

Sumerian Creation

The Sumerians developed the first of the great Mesopotamian civilizations in the third millennium B.C.E. Theirs was a mythology that stressed fertility and craftsmanship. Their culture influenced the later Semitic cultures in the Fertile Crescent (see also **Assyrian Creation; Babylonian Creation**).

Only fragments exist of the tablets with cuneiform script in which the Sumerian myths were written. We know enough, however, to be able to say that the goddess Nammu (Primeval Sea) gave birth to Heaven and Earth as a unified cosmic mountain of sorts. Heaven (An) was male, and Earth (Ki) was female (thus, Anki=universe or heaven-earth). They produced the air god, Enlil, who separated them so creation could move forward (see also **Creation from Division of Primordial Unity; Separation of Heaven and Earth**). Enlil carried his mother Ki downward, and An carried himself upwards.

Enlil lived in darkness, so he created the moon god, Nanna, who gave birth to the sun god, Utu. Then Enlil joined together with his mother, and it was this union of earth and air that made creation possible, beginning with the water god of vegetation and wisdom, En-ki or Enki (Ea in his later Semitic form), lord of the universe.

Some tablets seem to suggest that it was Enki who gathered the primeval waters into the Tigris and Euphrates rivers and who organized town life and the domestication of animals—indicated by phrases like "cities and hamlets" and "stalls and sheepfolds." Enki also stocked the swamps with fish, arranged the marshlands, and knew the ways of agriculture.

One of the Sumerian tablets says that Nammu (primeval sea) and the other gods wanted to create humans to serve them. At a somewhat drunken banquet, the gods tried to create these humans. Mother Earth made some that could not reproduce (see also **Imperfect Creation**), and Enki tried to form better ones out of clay (see also **Creation from Clay**). His humans could survive, but they were not particularly strong in body or spirit. So it is that human beings, the creations of drunken gods, have so many problems and weaknesses.

There came a time when the gods tired of humanity's failings and sent the Great Flood to destroy the world (see also **The Flood and Flood Hero**). Only Ziusudra (later Utnapishtim in the Semitic language of Babylon) survived with his wife in an ark of sorts. The influence on the Hebrew story of Noah's Ark is obvious.

The story of the Flood is preserved in the Babylonian Epic of Gilgamesh, a segment of which is retold here by Herbert Mason after the words of the flood hero himself. It is an epic that itself is based on an earlier Sumerian story of which we only have fragments.

> There was a city called Shurrupak
> On the bank of the Euphrates.
> It was very old and so many were the gods
> Within it. They converged in their complex hearts

Gilgamesh, the Babylonian hero, holds a lion in his arms. Only fragments remain of the Sumerian story on which the Babylonians based the Epic of Gilgamesh.

On the idea of creating a great flood.
There was Anu, their aging and weak-minded father,
The military Enlil, his adviser,
Ishtar, the sensation craving one,
And all the rest. Ea, who was present
At their council, came to my house
And, frightened by the violent winds that filled the
air,
Echoed all that they were planning and had said.
Man of Shurrupak, he said, tear down your house
And build a ship. Abandon your possessions
And the works that you find beautiful and crave
And save your life instead. Into the ship
Bring the seed of all the living creatures.

I was overawed, perplexed,
And finally downcast. I agreed to do
As Ea said but I protested: What shall I say
To the city, the people, the leaders?

Tell them, Ea said, you have learned that Enlil
The war god despises you and will not
Give you access to the city anymore.
Tell them for this Ea will bring the rains.

That is the way gods think, he laughed. His tone
Of savage irony frightened Gilgamesh
Yet gave him pleasure, being his friend.
They only know how to compete or echo.

But who am I to talk? He sighed as if
Disgusted with himself; I did as he
Commanded me to do. I spoke to them
And some came out to help me build the ship
Of seven stories each with nine chambers.
The boat was cube in shape, and sound; it held
The food and wine and precious minerals
And seed of living animals we put
In it. My family then moved inside
And all who wanted to be with us there:
The game of the field, the goats of the Steppe,
The craftsmen of the city came, a navigator
Came. And then Ea ordered me to close

The door. The time of the great rains had come.
O there was ample warning, yes, my friend,
But it was terrifying still. Buildings
Blown by the winds for miles like desert brush.
People clung to branches of trees until
Roots gave way. New possessions, now debris,
Floated on the water with their special
Sterile vacancy. The riverbanks failed
To hold the water back. Even the gods
Cowered like dogs at what they had done.
Ishtar cried out like a woman at the height
Of labor: O how could I have wanted
To do this to my people! They were *hers,*
Notice. Even her sorrow was possessive.
Her spawn that she had killed too soon.
Old gods are terrible to look at when
They weep, all bloated like spoiled fish.
One wonders if they ever understand
That they have caused their grief. When the seventh
    day
Came, the flood subsided from its slaughter
Like hair drawn slowly back
From a tormented face.
I looked at the earth and all was silence.
Bodies lay like alewives dead
And in the clay. I fell down
On the ship's deck and wept. Why? Why did they
Have to die! I couldn't understand. I asked
Unanswerable questions a child asks
When a parent dies—for nothing. Only slowly
Did I make myself believe—or hope—they
Might all be swept up in their fragments
Together
And made whole again
By some compassionate hand.
But my hand was too small
To do the gathering.
I have only known this feeling since
When I look out across the sea of death,
This pull inside against a littleness—myself—
Waiting for an upward gesture.

O the dove, the swallow and the raven
Found their land. The people left the ship.
But I for a long time could only stay inside.
I could not face the deaths I knew were there.
Then I received Enlil, for Ea had *chosen* me;
The war god touched my forehead; he blessed
My family and said:
Before this you were just a man, but now
You and your wife shall be like gods. You
Shall live in the distance at the rivers' mouth,
At the source.

Reprinted by permission of Houghton Mifflin Co. from Herbert Mason, *Gilgamesh: A Verse Narrative,* Boston: Houghton Mifflin Co., 1971.

Sources: Epic of Gilgamesh; Freund, 121–122; Kramer (B), 30–75; Kramer and Maier, 86–87.

Sumu Creation

These people of Central America believe that two brothers, the older of whom was Papan (Papa, Father) created the world *ex nihilo* (see also **Creation from Nothing**).

After they they made the beautiful world that we call nature, the brothers struck out in a canoe to admire their work and decide what they should do next. The rapids caught them, however, and they were thrown into the water. After swimming to shore, they made a fire to get warm and found some maize nearby to roast. When they threw the cobs onto the ground, animals sprang up and scampered away. Papan threw some of the cobs into the water and they became fish. Some he threw into the air and they became birds. Eventually there were animals everywhere and the brothers were well pleased.

Papan was so happy and entranced by a bird that, as he watched it, he stepped back into the fire and was consumed by the flames. He rose up into the air and became the sun. As the younger brother watched he also fell into the fire, but he struggled to stay on the earth; the sparks he made became the stars. Eventually he was consumed, however, and he became the moon.

The Sumu people are the children of Ma-Papan (Sun-Papa), conceived by his strong rays.
Source: Leach, 110.

Supreme Being

While most cultures have one god who is stronger than the others and is often considered their father, it the monotheistic cultures (see also **Christian Creation; Hebrew Creation; Muslim Creation**) that stress the absolute Supreme Being and make it the ultimate defining characteristic of their religions. It could be argued, too, that Brahman in Hinduism is clearly a Supreme Being, but for the fact that Brahman is unknowable and is more a concept than a being. In patriarchal cultures the Supreme Being or the head god is nearly always male, reflecting the realities of the family structure in human society. A goddess as Supreme Being almost always indicates a matrilineal culture or one firmly based in agriculture and, thus, the cycles of nature.

Swahili Creation

The Swahili-speaking people of Kenya and Tanzania have creation myths that are deeply influenced by the Islamic religion, which over the last centuries has gained strength in the area (see also **Muslim Creation**). Their story is an *ex nihilo* creation (see also **Creation from Nothing**).

There always was God; God is beyond birth and death. He creates by merely speaking (see also **Creation by Word**). So he said, "Let there be light," and, of course, there was light. God was so pleased with this light that he made some of it enter his prophet, Mohammed (see also **Christian Creation; Hebrew Creation**).

After God had made Mohammed's soul, he was so pleased with it that he decided to create humans so Mohammed could go to them as a prophet and teacher.

God knew everything that would ever happen, so he began creating things to fulfill the mysterious purpose that hid behind his knowledge. He made the Throne of Heaven—his throne, the Throne of the Last Judgment—and sat it upon the magnificent carpet he made, a carpet of every imaginable color that is huge beyond conception. Good souls live forever in joy in the shadow of the carpet under the Throne of Heaven.

Then God created the Mother of Books, a book with a soul of her own and one that is full of God's wisdom and the great secrets of life. God created the Pen with which to write down his commandments. The

Pen is huge beyond comprehension and has a conscious being. It has been writing down the history of mankind since it was given life by God.

Then God created the Trumpet and the archangel Serafili (Asrafel) to blow it at Doomsday.

God also made the Garden of Delights, a land of milk and honey, a paradise for good souls. He created the fire for the eternal punishment of souls who would not heed his messenger, Mohammed. He created the angels, the most important of whom was Jiburili (Gabriel), so full of God's radiance that even Mohammed fainted when he saw him in his original form. There were also the archangels Zeraili (Azrafel, the angel of death), Maliki (Michael, the angel of fire), and many others. God created the Cock of Heaven; his crow can be heard in the crowing of all the barnyard cocks who wake us each morning for the first prayers.

This is how God created the world: He rolled out the sky and placed the sun, moon, stars, and planets in it and set things to run according to the patterns we know. He made the universe of seven heavens, each with its own planet. These heavens are watched over by the souls of eight prophets: Adam, Isa (Jesus), Yahya (John the Baptist), Yusufu (Joseph), Idrisi (Enoch), Haruni (Aron), Musa (Moses), and in the highest heaven the ancient prophet Abraham, whose soul stands by the celestial mosque and its 70,000 praying angels.

Each heaven has its opposite hell; the farthest one away from the creator is for the unbelievers.

God spread out the carpet of earth, with its infinite variety of forms and shapes—seas, islands, mountains, valleys, trees, plants, deserts, insects, birds, fish, and animals. All the creatures have their own ways of praising God. God taught the creatures the laws of nature—how the big animals eat the small, and so forth.

God created time, too. Finally he called together the angels, and after the morning prayers, he spoke to them of his intention to create a being with their intelligence but made of clay. The angels had their doubts, and God let them speak. They saw the spectre of sin and war over the earth, but God reassured them; everything would work out according to his mysterious plan. The angels therefore agreed and sang God's praise.

Now God took clay and made Adam, whose name means earth (see also **Adam; Hebrew Creation**). He gave Adam life by speaking, "Life." Adam shuddered and came alive, and his first words were in praise of God.
Source: Sproul, 38–43.

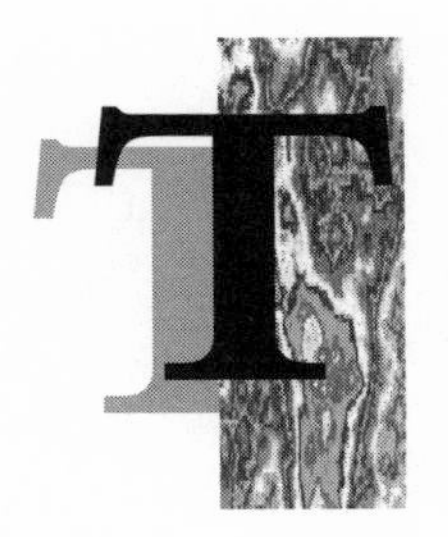

Tahitian Creation

Tahiti is part of the Society Islands in the South Pacific, and its creation stories contain the familiar Polynesian forms: Taaroa or Tangaroa, the cosmic egg, and creation *ex nihilo* (see also **Creation from a Cosmic Egg; Creation from Nothing; Maori Creation; Polynesian Creation; Society Islands Creation; Somoan Creation**).

For the Polynesians of Tahiti, Taaroa is the great god. He simply existed in space at the beginning of time before there was earth, sky, or humankind. In fact, he became the universe (see also **Animism**)—the sands, rocks, and light. Taaroa is the germ of life. He is within, under, above. The universe is his shell.

It is said that Taaroa lived in his egglike shell for a long time, but that finally he broke out of it and held it up to make the dome-sky, Rumia. Out of himself he made the world. His spine he used to make a mountain range, his ribs became hillsides, his fingernails became shells and scales, and his feathers became trees and plants. He kept his head to himself, however.

Taaroa made the gods and other things. Just as Taaroa had a shell, so does everything we know. The sky is a shell, earth is a shell for everything that lives in it, and the woman is the shell for human beings, because we are all born of woman. Taaroa is uniqueness itself. He is the rock in earth's center, out of which the world grew; he is the earth's surface.

He conjured the first man out of the earth, out of himself. This man was Ti'i. He created the first woman, too. Her name was Hina; she could see backwards and forwards and she was good. Ti'i was not good; he felt malice toward human beings.

Eventually there were wars in heaven among the gods and on earth among humans. This made Taaroa and his assistant, Tu (whom some say is the same as Ti'i), very angry. They cursed creation. Only the mitigation of the good Hina prevented the destruction of the world. Now, when storms come, Hina makes them finally go away, and when leaves fall, Hina makes new ones grow. Although Ti'i conjured up death for humans, Hina said she would bring them back to life. This is why the people say that men, not women, brought death.

Source: Sproul, 349–352.

Tantric Creation

Tantrism is an esoteric outgrowth of Hinduism. In Tantric doctrine, it is an energy source or emanation that is the creative germ. The slightest movement of this Absolute is felt throughout creation. The Tantric myth makes use of the Hindu creators—the self-existent **Brahman,** and even the specific personifications, such as Brahma and Vishnu, the creator and preserver gods (see also **Indian Creation**).

Tanzanian Creation

See Wapangwa Creation.

Tewa Creation

The Tewa language and Tewa traditions are peculiar to certain of the Rio Grande pueblos—San Juan, Santa Clara, San Ildefonso, and Nambe. The Tewa creation story, like those of most other Southwestern peoples, is an emergence myth (see also **Creation by Emergence; Keres Creation; Navajo Creation; Pueblo Creation**).

The people say that Long Sash, the evening star, once led the ancestors from the north to the place of the pueblos. They say that Long Sash was a great warrior and that he agreed to lead the people away from their marauding enemies. He warned them, however, that the journey would be hard. On the way, he taught them how to hunt and behave, and eventually they found a new country.

The creation story of the Tewa of the American Southwest, like that of the Navajo, is an emergence story. Artist Felipe Davalos depicts a version of the story in which the Tewa come from a lake in the north. Two women living in the lake ask a man to explore the region above the lake. Long Sash, or the Hunt Chief, returns to create Summer Chief and Winter Chief, at his right and left, to lead the Tewa from the lake.

Whereas it had been dark where they lived before, in the new country it was light all the time. The people walked around in this country, and they quarreled and fought until Long Sash made them stop. He ordered them to rest a while before deciding whether to continue following him or to go another way. After they were rested, the people did as they still do today when they have to decide something—they gazed at the two bright stars that lie north of Long Sash. After doing that they decided to follow Long Sash. He made sure they followed with good feeling toward each other.

After a while, everyone—including Long Sash—grew tired. Long Sash heard voices, and he fell into a long trancelike sleep to listen to them. When he awoke, he told the people he had been given signs. He said they would soon reach their proper destination and that, if they ever doubted, they should pray to the spirits above and look at his headdress for inspiration. He placed his headdress in the sky.

There are other ways of telling how the people got to the Middle World, where they live now. Many say, for instance, that it was **Spider Woman,** or Spider Grandmother, who was most helpful. They say that in the beginning there was only darkness and the people lived under the ground. The people became restless in the dark and began looking for another kind of life.

When Mole visited them from above, the people asked what it was like up there. Mole, of course, could not see, but he said it did feel different up there. He offered to lead the people up and said he would tell them when they were in a different world. As they followed Mole, they had to pile up behind them the dirt Mole dug up; therefore, they could never find the way back to the old world.

Finally they came out into a new world full of blinding light. The people were terrified and could not see. They covered their eyes and wanted to go back to the darkness, but then a little voice told them to wait and take their hands slowly away from their eyes. They did this, and there in front of them was Spider Woman, the old, stooped grandmother of everything that is.

Spider Woman was flanked by her twin grandsons, the War Twins. "Don't be foolish the way these boys are," she advised the people. "They are war-makers; don't waste your time fighting each other," she said. "To be happy, you must never use weapons."

Spider Woman pointed out the green growing thing, corn, to the people and told them how to work the land so they could grow it.

Then she pointed out the sacred mountains, and said the proper home for the people was near great Turtle Mountain (Sandia Mountain) in the south. She said that when the people found her again and their friend

Mole, they would have arrived at the right place. Then she faded into night. The people were terrified of the long night, and in the morning they ignored Spider Woman's advice and traveled to a mountain they could see clearly—to Red Mountain rather than to Turtle Mountain. There the Comanches killed many of the people; the mountain is called Los Sangres (meaning blood) for that reason. Then the people again disobeyed Spider Woman by quarreling, making weapons, and killing one another. It went on like this for a long time, and the war twins in the sky laughed at them.

They went back to the place of emergence but did not find Spider Woman there; they saw her in the sky in her beautiful web, shaking her head at their foolishness and weeping little star tears. Some of the people went up into the sky to the Grandmother.

Finally there were only two people left, a man and a woman. They took the hard road south, through the desert country. Then they saw some green trees and went over to them. There they found the beautiful Rio Grande.

A little turtle appeared in the sand. It was the turtle of Turtle Mountain; it had Spider Woman's sign on its back, and it left tracks like Mole tracks. The man and the woman knew that they had found the place to settle because they had found Mole and Spider Woman again. That is why the Tewa people live where they live now.

Source: Marriott (A), 80–95.

Theogony

Theogony is a work by Hesiod, a Greek poet and mythmaker of the eighth century B.C.E. It is the source of much of our knowledge of Greek mythology (see also **Greek Creation; Hesiod**).

Thinking Woman

Thinking Woman, also Thought Woman or Prophesying Woman (Tsitctinako or Sus'sistinako and various other spellings), is the creatrix of the Keresan pueblos of the Rio Grande Valley in New Mexico (see also **Acoma Creation; Keres Creation; Laguna Creation**). She is of the fertile underworld and projects her thoughts outward in creative acts of lifegiving.

Thompson, Stith

Stith Thompson is the compiler of *Motif-Index of World-Literature.* For students of the creation archetype, this is a useful source for patterns such as the ones also isolated by Maria Leach in *The Beginning: Creation Myths around the World* (see also **Leach, Maria**).

Thompson Indian Creation

This creation story of the Thompson Indians of British Columbia was told by an old shaman named Nkamtcine'lx at the turn of the century. The shaman said he had heard the myth from his grandfather. It is interesting that Old One's first *ex nihilo* creative act is to make five women (see also **Creation from Nothing**).

In the beginning of time there was only water everywhere. Old One got tired of looking at all the water, so he came down on a cloud determined to create something new. When the cloud—now fog—reached the waters, Old One plucked five hairs from his head (some say from his pubic area) and threw them down, and they became five perfect young women, already able to speak, see, and hear. Then he asked the women what they would like to do with their lives.

The first woman said she would like to have many children, be wicked, and pursue her own pleasure. She wanted her descendants to be fighters, murderers, adulterers, thieves, and liars. Old One was sorry for this answer.

The second woman said she too would like to bear children, but that she and her descendants would be good and true people—wise, honest, peaceful, and chaste. Old One praised the second woman and pointed out that in the end her way would triumph over the first woman's.

The third woman said she wanted to be the earth, the place where her sisters and their descendants would live. She would allow the people to take life from her, and she promised to give abundantly of herself. Old One was well pleased with new Earth Mother. He foresaw that she would nurture the world and then take the dead back to herself and keep them warm. She would give forth beautiful trees and plants.

The fourth woman said she planned to be fire, that she would give warmth to the people and help them make their food better. Old One was more than satisfied with this plan.

The fifth woman simply wished to be water.

Then Old One changed the women into their wishes for themselves. The third woman lay down in the waters and became the Earth

Mother on which we live. The fifth woman became the waters within Earth, the fourth woman became the spirit of fire in all things that burn. As for the first and second women, Old One placed them on earth and immediately impregnated them. "You will be the first people," he said, "and from you will come all the people of the earth—male and female." Old One foresaw that at first the evil woman's children would dominate but that eventually the good woman's children would prevail. Old One said he would bring together the five sisters and all of the people—good and evil, dead and alive—at the end of the world.

All of this explains why there are good and bad people on earth. It also explains how all of us are directly related to earth, fire, and water.
Source: Weigle, 187–190.

Thonga (Tonga) Creation

In this version of creation by separation, the Thonga people of the South Pacific say that in the time of darkness at the beginning, the gods Vatea and Tonga-iti fought over a child (see also **Creation from Division of Primordial Unity**). Each said the child belonged to him. Finally they agreed to cut the child in two. Vatea took the top half, squeezed it into a ball, and threw it into the sky, where it became the sun. Tonga-iti did the same with his half, but only after it had spent some days on the ground, bleeding. His half, therefore, became the somewhat pale and drained moon.
Source: Olcott, 30–31.

Tiamat as Creator

Tiamat was the primeval source of the Mesopotamian creations (see also **Babylonian Creation; Enuma Elish; Sumerian Creation**). She is sometimes the primeval sea, sometimes the formless matter of the universe. The patriarchal Babylonians claimed that Tiamat was cut in half into the heavens and earth (see also **Creation from Division of Primordial Unity; Separation of Heaven and Earth**) by her son, the great god Marduk. In earlier times she would have accomplished the division herself.

Tidal Creation Theory

The tidal creation theory was developed by English scientist James Jeans early in the twentieth century. It suggests that the gravitational pull of

a stray star caused huge tides on the sun's surface and that the crest of a particularly high wave was pulled completely away. It scattered, like wave spray, into space. These bits of spray are the planets—including ours. Earth was at first a fiery chunk of spray that after several billion years of cooling formed a crust (see also **Creation in Science**).
Source: Leach, 18–19.

Tierra del Fuego Creation

The tribes of Tierra del Fuego—the Ona, Yahgan, and Alacaluf Indians—live at the southern tip of the world, the islands off Cape Horn. They believe in a supreme being called Temaukl or Xelas. His attributes are many, and his names mean things like One Above, Old One, Father, Good One, Murderer in the Sky, Star, and so forth. Most of the people in this region think of the supreme god as also the creator.

Before anything else Temaukl—the forever-existing—was. He made the heavens, earth, and people. He gave life and he gave death, and he still does.

It was Kenos, the first man, who was sent to the world to bring order to it. It was he who made the plants and animals and parceled out the land to the people. The world was and still is a difficult place because of the constant north-south struggle between a bit of warmth and the desperate cold. The people always leave the god a bit of food or a hot coal in hopes of better weather.

The Yahgans say the god's name is Watauinewa and that he did not actually create the world, but that he did give people life.
Source: Leach, 129–131.

Trickster

Tricksters appear in many cultures. Hermes in Greece is a trickster of sorts, and so is Brer Rabbit in the folklore of the American South. The figure is especially popular among Native American and African peoples. He usually plays an important role in creation (see also **Coyote; Raven**) because he represents the amoral, creative power that exists in the preconscious stages of human development. Often he uses his wiles to steal things, such as fire, for humans. He has what might be called the creative power of dream. Thus, the trickster is often highly erotic and apparently immoral. He is almost always funny and is frequently the butt of his own tricks (see also **Eskimo Creation; Inktomi; Krachi Creation; Sioux Creation, Spiders in Creation**).

Truk Island Creation

Truk Island is one of the Carolines in Micronesia, the group of small islands in the South Pacific that includes the Gilberts, the Marshalls, and the Marianas (see also **Gilbert Islands Creation; Mariana Islands Creation; Marshall Islands Creation**).

The people of Truk Island say the sky god's daughter, Ligoububfanu, was the mother of humans, animals, coconuts, and grain. It is said that her first child's face is stamped on the coconut.
Source: Leach, 182–183.

Tsimshian Creation

The Tsimshians are a tribe who live at the southern end of Queen Charlotte Island off British Columbia. Their creation myths are mostly of the **Raven** cycle (see also **Trickster**). By hiding within the big chief's daughter, Raven accomplishes a version of the **earth-diver creation.**

When the world was still dark, there was a chief whose son got sick and died. The parents mourned their son and made everyone else do so. One morning the mother went to look at her son's body and was surprised to find a dazzling young man there. He had been sent by the sun, he said, to stop the wailing of the parents and their people. The parents thought this was their son who had returned.

The young man, despite the urging of the mother, would not eat until he met two slaves, one male and one female, called Mouth at Both Ends. After meeting them and being affected by their power, he ate everything there was to eat in the settlement. The chief gave him a Raven's mask and renamed him Giant. The chief also gave Raven-Giant a bladder full of seeds and suggested that he fly to the mainland and plant them for food. He also gave him a stone to rest on in case he got tired.

Raven-Giant stole daylight so the world could be light and so his plants would grow. He did this by flying up to heaven and hiding in the drinking bucket of the big chief's daughter there. When she drank, she drank him and became pregnant. Then he was born in the sky chief's house and was able to steal some light. Raven-Giant brought all sorts of things to humans—including death.

There are other Tsimshian myths of the coming of light. This one was told to anthropologist Franz Boas by a Tsimshian Indian in 1916.

In the beginning, before anything that lives in our world was created, there was only the chief in the sky. The chief had two sons and a daughter, and his people were numerous. But there was no light in the sky—only emptiness and darkness.

The chief's eldest son was named Walking-About-Early, the second son was called The-One-Who-Walks-All-Over-the-Sky, and the daughter was Support-of-Sun. They were all very strong, but the younger boy was wiser and abler than the elder.

It made the younger son sad to see the sky always so dark, and one day he took his brother and went to cut some good pitch wood. They bent a slender cedar twig into a ring the size of a person's face, then tied the pitch wood all around it so that it looked like a mask. They lit the wood, and The-One-Who-Walks-All-Over-the-Sky put on the mask and went to the east.

Suddenly everyone saw a great light rising. As the people watched and marveled, the chief's younger son ran from east to west, moving swiftly so that the flaming mask would not burn him.

Every day the second son repeated his race and lit up the sky. Then the whole tribe assembled and sat down to a council. "We're glad your child has given us light," they told the chief. "But he's too quick; he ought to slow down a little so we can enjoy the light longer."

The chief told his son what the people had said, but Walks-All-Over-the-Sky replied that the mask would burn up before he reached the west. He continued to run very fast, and the people continued to wish he would go slower, until the sister said, "I'll try and hold him back a little."

The next time Walks-All-Over-the-Sky rose in the east and started on his journey, Support-of-Sun also started from the south. "Wait for me!" she cried, running as hard as she could. She intercepted her brother in the middle of his race and held him briefly until he could break free. That's why the sun today always stops for a little while in the middle of the sky. The people shouted for joy, and Support-of-Sun's father blessed her.

But the chief was displeased with Walking-About-Early because he was not as smart and capable as his younger brother. The father expressed his disappointment, and Walking-About-Early was so mortified that he flung himself down and cried. Meanwhile his brother, the sun, came back tired from his daily trip and lay down to rest. Later when everybody was asleep, Walking-About-Early rubbed fat and charcoal over his face. He woke his little slave and said, "When you see me rising in the east, jump up and shout, 'Hurrah! He has arisen!' "

Then Walking-About-Early left, while Walks-All-Over-the-Sky slept deeply, his face shedding light out of the smoke hole. Suddenly Walking-About-Early rose in the east, and his charcoaled face reflected the smoke hole's luster. The little slave jumped up and shouted, "Hurrah! He has arisen!"

Several people asked him "Why are you so noisy, bad slave?" The slave jumped up and down, pointing to the east. The people looked up and saw the rising moon, and they too shouted, "Hurrah!"

Time passed, and animals were created to live in our world below. At last all the animals assembled to hold a council. They agreed that the sun should run from east to west, that he should be the light of day, and that he should make everything grow. The moon, they decided, should walk at night. Then they had to set the number of days that would be in a month. The dogs were wiser than the other animals and spoke first. "The moon shall rise for forty days," they said.

All the animals were silent. The dogs sat together talking secretly among themselves and thinking about what they had said. The wisest dog, their spokesman, was still standing. He was counting up to forty on his fingers, when the porcupine suddenly struck him on the thumb. "Who can live if there are forty days to each month?" the porcupine said. "The year would be far too long. There should be only thirty days in a month."

The rest of the animals agreed with the porcupine. And as a result of this council, each month has thirty days and there are twelve months in a year. By now the animals were disgusted with the dogs and banded together to drive them away. For this reason dogs hate all the creatures of the woods, and most of all the porcupine, who struck the wise dog's thumb with its spiny tail and humiliated him in the council. And because of the porcupine's blow, a dog's thumb now stands opposite to his other fingers.

Before that long-ago council ended, the animals also named the following months:

Between October and November, Falling-Leaf Month
Between November and December, Taboo Month
Between December and January, The Intervening Month
Between January and February, Spring Salmon Month
Between February and March, Month When Olachen Is Eaten
Between March and April, When Olachen Is Cooked
Between May and June, Egg Month
Between June and July, Salmon Month
Between July and August, Humpback-Salmon Month
Between September and October, Spinning Top Month

In addition, the animals divided the year into four seasons—spring, summer, autumn, and winter.

New things were also happening in the sky. When Walks-All-Over-the-Sky was asleep, the sparks that flew out of his mouth became the stars. And sometimes when he was glad, he painted his face with his sister's red ochre, and then people knew what kind of weather was coming. If his red paint colored the sky in the evening, there would be good weather the next day, but a red sky in the morning meant that storms were coming. And that's still true, people say.

After the sky had been furnished with the sun, moon, and stars, the chief's daughter, Support-of-Sun, was cast down because she had played such a small part in the creation. Sadly she wandered westward into the water, and her clothes became wet. When she returned, she stood near her father's great fire to warm herself. She wrung the water out of her garments and let it drip onto the flames, making a great cloud of steam that floated out of the house. It settled over the land and moderated the hot weather with damp fog. Her father blessed her, for the whole tribe enjoyed it. And to this day, all fog comes from the west.

The chief was glad when he saw that all three of his children were wise. Now it was the duty of the moon, Walking-About-Early, to rise and set every thirty days so that people may know the year. The sun, Walks-All-Over-the-Sky, was charged with creating all good things, such as fruit, and making everything plentiful. And the chief's daughter,

Support-of-Sun, served by refreshing the hot earth with cool fog.

Source: Bierhorst, 28–29.

Tuamotuan Creation

The myths of the Tuamotuans of the South Pacific resemble those of the Maoris and other Polynesians. There are also elements of the Indian creation with its sleeping gods and its sense of time breaking into the timelessness of Brahman. The Tuamotu creator god is Kiho, who created by the power of his words and thoughts from nothingness (see also **Creation by Thought; Creation by Word; Creation from Nothing**).

Kiho lived alone in the emptiness under Havaiki, or nonland. He had no parents, no mate. His only company was his double, his Activating Self. He thought within himself and acted through the Activating Self.

Kiho evolved the dark waters first and then the night world of the spirits. Then came the day world of our earth and finally the sky world.

This is how it happened: Kiho awoke and gazed into the immemorial chaos. Then he spoke what it was— "the total darkness of Havaiki."

Kiho thought of things and called on his Activating Self to give concreteness to his inner knowledge. Then the creative urge within the primeval waters and the land began to live. Beneath Havaiki, sleeping fruitfulness, sleeping sky, and sleeping land began to awaken.

Then Kiho gave utterance to the primordial waters and they began to be real. He called on his Activating Self to bring earthquakes to the rock foundations of Havaiki. He made his eyes into flames and there was light. Kiho floated up from the depths as his Activating Self and lay on the waters. He created the night world and the day world. Gradually Kiho raised himself and created the heavens and the earth. Next he made Atea-ragi, the male force, and Fakahotuhenua, the female force, the "fructifier of the soil."

Kiho turned now to organizing the world—placing the sand and sea in their proper places—and he organized the heavens. The Activating Self of Kiho was incarnated to become the ruler of the world, and Kiho drifted back to the non-being.

The following is a cosmic chant from the Takaroa Atoll of the Tuamotu Islands, as translated by Kenneth Emory:

> Life appears in the world,
> Life springs up in Havaiki.
> The Source-of-night sleeps below
> in the void of the world,
> in the taking form of the world,
> in the growth of the world,
> the life of the world,
> the leafing of the world,
> the unfolding of the world,
> the darkening of the world,
> the branching of the world,
> the bending down of the world.

Reprinted from Kenneth Emory, trans., "Cosmic Chant from the Takaroa Atoll of the Tuamotus," *Journal of the Polynesian Society,* vol. 47, no. 5 (1920): 14.

Source: Sproul, 352–358.

Tungus Creation

The Tungus are a shamanistic people of Siberia. As in so many Siberian myths, the devil plays an active, if negative, role in creation. This **creation from chaos** myth is really a reflection of the struggle between witchcraft and true shamanism. Here we also find the world tree motif as in Icelandic mythology (see also **Yggdrasil**).

In the beginning there was only the waters until God sent down fire. The fire burnt part of the ocean, and land came. As soon as God set foot on the new land he found Buninka, the devil, there. Buninka wanted to create his own world, and he got so angry when God refused to let him do so that he broke God's lyre.

God challenged Buninka to make a tree out of the water to prove his powers. Buninka agreed, but the pine tree he created was weak and it leaned over. God created a tree that was strong and constantly growing. Buninka therefore recognized God's power over him and all things.

Source: Sproul, 217–218.

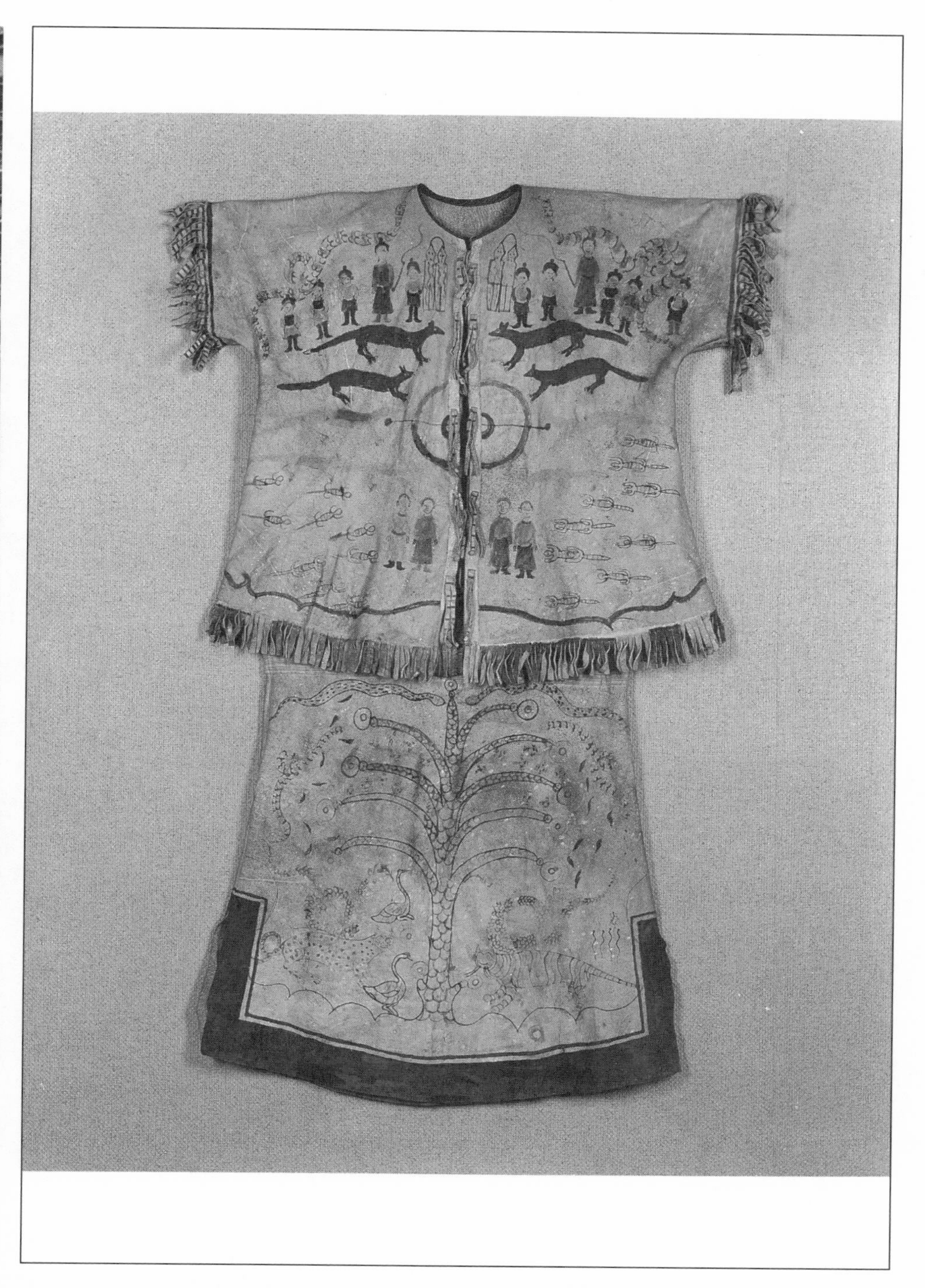

An elaborately decorated Tungus shaman's garment from Northeast Asia—Siberia—has a tree on its skirt. According to Tungus tradition God, challenged by the Devil, made a tree that grew constantly while the Devil's tree was weak and bent over. Such trees usually represent a separation and layering between heaven and earth.

Turtle as Earth-Diver

Turtle is one of the most popular of earth-diver animals. His back is often the foundation of the newly created world (see also **Earth-Diver Creation; Huron Creation; Maidu Creation**).

Tylor, Sir Edward

Evolutionist and anthropologist Sir Edward Tylor, who developed the concept of **animism,** is the author of *Primitive Culture* (1871).

Uitoto Creation

The creation myth of the Uitoto Indians of Colombia is unusual in that it postulates the illusion of reality before the actual existence of reality. This is the epitome of creation *ex nihilo* (see also **Creation from Nothing**).

First there was only a vision, an illusion that affected Nainema, who was himself the illusion. Nothing else existed. Nainema took the illusion to himself and fell into thought. He held the vision by the thread of a dream and searched it, but he found nothing. Then he searched it again, and he tied the emptiness to the dream thread with magical glue. Then he took the bottom of the phantasm and stamped on it until he could sit down upon this earth of which he had dreamed (see also **Creation by Thought; The Dreaming**).

As he held onto the illusion, he spat out saliva (see also **Creation by Secretion**) and the forests grew. He lay down on the earth and made a sky above it.

Gazing at himself, the One who was the story created his story for us to hear.

Sources: Eliade (B), 85; Weigle, 181–182.

Upanishads

The Upanishads are a series of late Vedic writings on the nature of the universe. They are a source for some of the **Indian creation** stories.

Ute Creation

The Ute Indians of southwestern Colorado tell a **creation from chaos** myth that reflects their sense of the need to live in harmony with all of nature.

In the beginning Manitou, or Great He-She Spirit, lived alone in the sky. There were only sky, clouds, sun, and rain with the spirit. Manitou decided to make something different, so He-She drilled a hole in the sky to look down to the vast emptiness. He-She poured rain and snow into the hole, then took the dirt and stones from the hole and poured that through, too.

Later, when Manitou looked down, there was a mountain where he had poured the dirt; there were lots of other mountains, too, and a great plain. He-She stepped down onto the big mountain and immediately improved it by producing trees and plants with a mere touch. Manitou improved the plains by waving his hands over them, bringing sweet grasses. He-She caused the sun to shine through the hole made earlier; this melted the snow, bringing streams and rivers that flowed into the oceans, the sky blue waters, who stole their color from the sky.

Then came rain and the fruition of the earth, and He-She continued making things. The broken-off ends of his cane were made into fish. He-She had to breathe on them to give them life. The Utes used to eat fish, but they don't anymore because once some wicked people threw their dead victims into the water, and it is impossible to tell the right fish from the fish who used to be the dead people.

Manitou also made birds out of the beautiful leaves in the forest. Eagles came from oak leaves, hawks from red sumac, and so forth. He-She made animals from the middle of his cane. The animals lived in harmony until **Coyote** (see also **Trickster**) caused trouble and the animals began to fight.

Manitou decided to make a boss animal who would rule wisely and keep things in order. This is why he made Grizzly Bear, who is still chief among animals. After creating the bear and establishing laws of behavior, Manitou left for the heavens.

Source: Wood, introduction.

Wahungwe Creation

The mysterious **creation from chaos** myth of the Wahungwe Makoni people of Rhodesia makes clear an essential truth of life—that with fertility and life comes death; life is, in effect, death-defined.

In the beginning Maori created the first man, Mwuetsi, who was the moon. He placed him at the bottom of the waters, Dsivoa, with a ngona horn filled with ngona oil, but Mwuetsi insisted that he wanted to live on earth. Maori did not approve, but he placed Mwuetsi on earth, saying his living there would end in his death.

On earth Mwuetsi found only barrenness, and he complained to Maori. "I warned you," said Maori, but he sent the man a woman to keep him company. This woman was Massassi, the morning star, to whom Maori gave the gift of fire. Maori said Mwuetsi could keep Massassi with him for two years.

The first couple went at night into their cave. Mwuetsi gathered kindling and Massassi made fire by twirling the firemaker. Then the man lay on one side of the fire and the woman on the other. Mwuetsi lay awake wondering why God had sent him the maiden. Finally, he took the ngona horn he had been given and moistened his finger with some of the oil. He leapt over the fire and touched Massassi with his oiled finger. Then he went back to his side and slept.

In the morning Massassi was huge, and soon she gave birth to plants and trees until the whole earth was covered by them.

The couple lived well and happily; they learned to build, to trap, and to grow vegetables. Mwuetsi worked on gathering wood and water, Massassi cooked. At the end of the two years, however, Maori took Massassi away, and Mwuetsi wept for eight years.

Then Maori reminded Mwuetsi of his original warnings against coming to earth. He sent another woman, Morongo, the Evening Star, however, saying she also could stay two years. When on the first night Mwuetsi touched Morongo with his oiled finger, she told him she was not like Massassi, and that he would have to oil their loins and then have intercourse with her. This Mwuetsi did that night and each night after that. Every morning Morongo gave birth to the animals of creation. On the fourth morning she gave birth to human boys and girls, who were grown by nighttime.

On the fourth night Maori sent a terrible storm and again warned Mwuetsi that he was heading toward his death with all of this procreation. Morongo, now a temptress, told Mwuetsi to build a door so Maori could not see what they were doing, and the man and woman continued to sleep together against God's command (see also **Eve; Fall of Humankind**).

In the morning Morongo gave birth to the violent animals—the lions, leopards, scorpions, and snakes.

On the fifth night Morongo told Mwuetsi to have intercourse with his daughters, and he did (see also **Incest in Creation Myths**). His daughters bore children and became the mothers of the people. Mwuetsi became the Mambo—great chief—of his people. He lived in Zimbabwe, the royal precinct.

One night Morongo coupled with a snake (see also **Eve**), and when one day Mwuetsi wished to sleep with her she was reluctant. Mwuetsi insisted, and in the night his wife's snake-lover bit him and he became ill.

As his sickness increased, the rivers and the fruits of the earth dried up. Even the animals and people died. The sacred dice said that only if Mwuetsi were sent back to the depths would things become better, so the children of Mwuetsi strangled him and buried him. They also buried Morongo with him. The two years were up, and death had come as Maori had said it would.

Each morning the Mwuetsi, the moon, rises from the sea and follows his beloved first wife, Massassi, the morning star, across the heavens.
Sources: Beier, 15–17; Freund, 147–152.

Wakaranga Creation

See Wahungwe Creation.

Wapangwa Creation

The Wapangwa people of Tanzania tell this *ex nihilo* story, which makes use of the theme of excretory creation (**Creation by Secretion; Creation from Nothing**).

Before there was a sun, a moon, or stars, there was only the wind and a tree (see also **Yggdrasil**) on which there lived some ants. There was also the Word, which contolled everything, but the Word could not be seen (see also **Christian Creation**); it was a catalyst for creation.

Once the wind became angry at the tree for standing in its way, so it blew particularly hard, tearing off a branch on which there were white ants. When they landed, the ants were hungry, so they ate all of the leaves on the branch, sparing only one, on which they defecated a huge pile.

Then they had no choice but to eat their own excrement, and over time, as they ate and redeposited their excrement, the pile became a mountain that finally spread to the original tree. By then the ants preferred excrement to leaves, and they continued the process of adding to the pile until it had become the earth.

The wind still blew on the world so strongly that parts of the excrement pile began to harden into stone. The world gradually formed, until the Word sent snow and then warm wind, which melted the snow and brought a huge flood (see also **The Flood and Flood Hero**). The waters killed the ants; there was water covering everything.

Later the earth and the world tree joined, and the trees, grasses, rivers, and oceans took form. The air gave birth to beings that flew about singing. These beings came to earth and became animals, birds, and humans, each with its own song or language.

The new beings were hungry. The animals wanted to eat the Tree of Life, but the humans defended it. This led to a huge war between humans and animals and to the tradition of humans and animals eating each other. The war was so ferocious that the earth shook, and bits and pieces of it flew off, gained heat, and became the sun, moon, and stars.

After the war there was the creation of gods, rain, thunder, and lightning. A long-tailed sheep with a single horn was so happy at the end of the great war that she leapt into the air, caught fire, and became the source of thunder and lightning.

The new gods who sprang up were harsh with humans. One of them told the people that the sheep that had sprung into the air had killed the Word, the ultimate creator, and that the people would be reduced in size and in the end would be consumed by fire.

Source: Beier, 42–46.

Winnebago Creation

There are several versions of the creation story told by the Winnebago Indians of the upper Midwest, but all are basically of the *ex nihilo* type (see also **Creation from Nothing**). This retelling is based on a combination of several myths.

Things began when Earthmaker came to consciousness and realized he was alone. Taking a bit of whatever it was he was on, he made a ball and sent it flying into space as the world.

He made hair for the naked earth out of grass, but the earth did not stop spinning. He thought of a tree (see also **Creation by Thought**), made one, and threw it down, and other trees grew up. He made the four directions and made rocks that became mountains. These stopped the wild spinning, and the sun came out.

Earthmaker then created the birds, land animals, and fish.

He thought next of humankind, making these weak beings last. He did it by shaping a bit of earth (see also **Creation from Clay**) in his own image. The man could neither see, hear, nor speak. Earthmaker put a finger into his own ear and then into the human's ear, and then the man could hear. He did the same between his eyes and the man's and between his mouth and the man's, and then the man could also see and speak.

The man did not seem to know what to say, however, and Earthmaker understood that he needed something more to make him whole. So it was that he breathed life into the man with his own breath so he and the man could converse.

Finally Earthmaker sent the man into the world and split him into the many peoples who live here now.

Source: Leach, 79–81.

Wisdom

In Chapter 8 of the book of Proverbs in the Hebrew Bible, we find the female figure, Wisdom, who is said to have been a partner with Yahweh in the creation. Wisdom in Greek is **Sophia,** but in Christian theology, Sophia was gradually supplanted as the mediator between the creator and his world by the male Logos, the Word (see also **Christian Creation**).

In Proverbs we hear Wisdom say, "I was created by the Lord before he made anything else, before he made the earth." Wisdom reveals that during the creation she was at the Lord's side, "his darling and his delight . . . while my delight was in mankind"(see also **Hebrew Creation**).

World Parent Creation

The Earth Mother-Sky Father motif is the basis for world-parent creations. In this type of myth, which can be seen in the Fon, Polynesian,

Zuni, Egyptian, Minyong, Yoruba, Yuma, and Greek creations, among others, the universe evolves from a union of the opposite qualities of sky and earth, the heights of thought and the depths of matter. Usually creative space between the world parents is made—sometimes violently—by their own children (see also **Creation from Division of Primordial Unity; Enuma Elish; Separation of Heaven and Earth**).

World Tree

This is the axle tree, the unifier and separator of the world parents, earth and sky. It is the created world in microcosm (see also **Yggdrasil**).

Wulamba Creation

See Djanggawul Creation.

Wyandot Creation

The Wyandot people of Kansas and Oklahoma (originally of the Great Lakes region) are related to the Hurons. Not surprisingly, their principle creation myth involves the earth-diver Star Woman motif that exists among most Iroquoian tribes and their relatives (see also **Cherokee Creation; Earth-Diver Creation; Huron Creation; Iroquoian Creation**).

As in the myths of all of these cultures, Turtle plays an important role in the Wyandot creation. The people say that Little Turtle was the creator of the sun—a council of the animals ordered him to create it. It is said that he married the sun to the moon and that the frolicking stars are their children.
Source: Olcott, 8.

Wyot Creation

Like so many California tribes, the Wyots tell a story about an **imperfect creation** and a flood (see also **The Flood and Flood Hero**). There is a clear indication in this *ex nihilo* myth of the influence of Christian missionaries. Note, for instance, the basket/ark.

Old Man in the heavens created people, but they turned out all furry, and he decided to get rid of them with a flood. Condor somehow knew

about this and made a basket into which he got with his sister. Soon they were floating, and after a time they made a hole in the basket and looked out. They saw land, animals, birds, and things, but no more furry people. Condor decided to mate with his sister (see also **Incest in Creation Myths**), and the first real people were born. They looked just right, and they made more people. Old Man in the heavens was happy.
Source: Sproul, 236.

Yahweh

Yahweh is a name for the Hebrew creator, who is a patriarchal and monotheistic deity of the Semitic peoples. He formed part of the Indo-European invasion into the Fertile Crescent region (dominated by Great Goddess religions in ancient times. He is one of the early versions of what in monotheistic religions—especially Judaism, Christianity, and Islam—is thought of as God (see also **Hebrew Creation**).

Yakima Creation

The Yakima are mountain people of eastern Washington. Their *ex nihilo* creation story has an earth-diver aspect (see also **Creation from Nothing; Earth-Diver Creation**).

First there was water everywhere and only the great sky chief, Whe-me-me-ow-ah, to see it. The chief went to where the water was shallow and brought up handfuls of mud that turned into land. Some of the piles were high enough to freeze and become mountains.

Whe-me-me-ow-ah made trees grow and also fruits and grasses. He made a man out of a bit of the mud (see also **Creation from Clay**). He taught the man how to hunt and fish. The man was missing something, so Whe-me-me-ow-ah made a woman to be with him. He taught her how to make baskets and how to collect and prepare food.

One night, the woman had a dream in which Whe-me-me-ow-ah gave her a quality that no one could see or touch but that women pass on to their daughters.

The people began fighting, even though the world they had was good. Mother Earth became so angry at them that she shook herself violently and many people died under falling rocks and mountains (see also **The Flood and Flood Hero**). These people still live as spirits on the mountaintops. They can be heard wailing and moaning all the time. It is said that a time will come when the Great Spirit will uncover the bones of the destroyed people and the spirits will return to those who in life were true followers of the elders' beliefs.
Source: Erdoes, 117–118.

Yami Creation

The Yami are an Indonesian people who live on an island off the tip of Taiwan (Formosa) called Botel-Tobago. This is their creation story in which a stone and a plant become cosmic eggs of sorts (see also **Creation from a Cosmic Egg**).

The Creator looked down from the heavens and liked the island of the Yami, and he dropped a stone on the place that would become Ipaptok Village (a *paptok* is a bean plant from which the first people got food). When the stone hit the ground, a man came out, and he ate of the plant. As he walked about later, he saw a piece of bamboo growing on the shore. It divided, and a man stepped out. One of the men asked the other, "Who are we?" and the other said, "We are man."

Bamboo son walked one way, and found silver, while Stone son found iron. The two sons came home and worked on the iron and silver—one hard, one soft. One day a boy child burst from the swollen right knee of Bamboo son and a girl sprang from his left knee. The same thing happened to Stone son. The children grew, and in time they married properly: Bamboo with Stone, Stone with Bamboo. Everyone was happy.

Then the people made canoes. Bamboo son's silver axe was too soft to fell trees, so the people made beautiful helmets out of the silver and kept the iron for axes. The canoes were beautifully designed, but Bamboo son put the support ribs on the inside of his canoe and it broke up at sea. Stone son's ribs were on the outside, and his canoe survived, but it leaked. He fixed the leak with the fiber of the kulau tree, which the people still use today to make plugs for stopping leaks.

In their silver helmets and outer-ribbed canoes, the people celebrate the sacred fish calling at the Flying Fish Festival by singing of Ipaptok, where man first burst out of the dark nothingness.
Sources: Del Rey; Leach, 159–163.

Yana Creation

The Yana people of California tell this story. It is more an origin story than a true creation myth, since it explains the phases of the moon.

A boy called Pun Miaupa fought with his father and ran away from home to his uncle's house. He announced to his uncle his desire to win Halai Auna (Morning Star) as his wife. His uncle, however, knowing the cruelty of Wakara (Moon, the girl's father), tried to talk his nephew out of the idea, but the boy insisted and left for Wakara's house. The uncle, a shaman, entered his nephew's heart to protect him.

Although Wakara was not outwardly impolite, he set to work immediately to kill the boy by magically transporting his family group to the house of Tuina, the sun. There he assumed that Pun Miaupa would die from the sun's poisoned tobacco. The shaman-uncle in the boy's heart saved him, however, and the uncle caused a great flood (see also **The Flood and Flood Hero**) to drown everyone in Wakara's and Tuina's families except for Halai Auna. It was only because of the girl's unhappiness at the loss of her family that the magician relented and brought the evil ones back to life.

Everyone returned to the home of Wakara, where he proposed a treebending contest. Because of the uncle's magic, Wakara lost and was flung into the sky, where he remains today. Pun Miaupa and Halai Auna married, and the boy shouted out to Wakara in the sky that he was doomed to stay where he was, growing old, then young, and then old again, forever and ever. Thus the moon has its phases.

Source: Olcott, 21–23.

Yao Creation

The Yao are a Bantu tribe of Mozambique. Their creation myths resemble those of other African peoples in that they stress the negative role of humankind in destroying the harmony of the natural world, going so far as to suggest that humans chased the gods away. Some of the Yao people say they emerged—birthlike—from a hole in the earth (see also **Creation by Emergence**). Others say they were taken from the waters. It is of interest to note that, as in so many Native American creations, the creator of this African myth is aided by an animal.

In the beginning there was only the creator, Mulungu, and the animals until Chameleon found a tiny man and a tiny woman in his otherwise empty fish net. He took the net and strange creatures to Mulungu, who instructed Chameleon to let them go so they could grow—and grow they did.

The new people learned to make fire, and often they terrified the animals by setting the forest on fire. They learned to hunt, and they killed and ate buffalo and other animals. As for the animals, they learned to flee in horror from humans. Chameleon, for instance, fled into the trees. Mulungu was so disgusted that he asked the spider to spin a rope so he could escape to the heavens. Mulungu and all of the gods stay away now.
Source: Leach, 143.

Yaruro Creation

The Yaruros are Venezuelan Indians whose social arrangements are matrilineal. To this day, the high deity of these people is the Great Mother, Kuma, whom shamans visit in their ritual trances.

In the beginning there was nothing. Puana, the water serpent, came and created the world *ex nihilo* (see also **Creation from Nothing**). His brother Itciai, the jaguar, created water. It was their sister, Kuma, who made the Yaruro people. These people became members of the Water Serpent clan or the Jaguar clan. It was only later that other people were created, people like the white Racionales.

The Yaruros were given horses, but they were frightened by their size and gave them to the Racionales, who still ride them today while the Yaruros walk or canoe.

At night the Sun, who travels by canoe across the sky, goes to visit his wife, Kuma. The stars are their children. The Sun's sister, the Moon, travels in a larger boat.

Kuma gave the Yaruros good plants, but they cut them from the top so the seeds fell onto the neighboring land of the Racionales. Now the Yaruros have simple fruits while their neighbors have better things, like maize, tobacco, plantains, and bananas.

Several versions of the Yaruro myth, as retold by Vincenzo Petrullo, follow (see also **Earth-Diver Creation**).

1. Everything sprang from Kuma, and everything that the Yaruros do was established by her. She is dressed like a shaman, only her ornaments are of gold and much more beautiful.

   With Kuma sprang Puana and Itciai; Hatchawa is her grandson and Puana made a bow and arrow for him. Puana taught Hatchawa to hunt and fish. When Hatchawa saw the people at the bottom of a hole and wished to bring them to the top Puana made him a rope and a hook.

Another figure that sprang with Kuma was Kiberoh. She carried fire in her breast and at Kuma's request gave it to the boy Hatchawa. But when the boy wanted to give it to the people Kuma refused and he cleverly threw live fish in the fire, spreading coals all about. The people seized the hot coals and ran away to start fires of their own. Everything was at first made and given to the boy and he passed it on to the people. Everybody sprang from Kuma, but she was not made pregnant in the ordinary way. It was not necessary.

2. The first to appear was Kuma, the chief of all of us and the entire world. Itciai, Puana, and Kiberoh appeared with her. There was nothing then. Nothing had been created. Kuma was made pregnant. She wanted to be impregnated in the thumb but Puana told her that too much progeny would be produced that way. So she was made pregnant in the ordinary way. Hatchawa was born, grandchild (?) of Kuma, Puana, and Itciai. From then on the attention of the three was centered on the boy. Puana created the land; Itciai the water in the rivers. Hatchawa was very small, but soon grew to a very large size. Kuma and Puana took care of his education, though Puana took more care of him. Puana made a bow and arrow for him and told him to hunt and fish. Hatchawa found a hole in the ground one day and looked into it. He saw many people. He went back to his grandparents to ask them to get some of the people out. Kuma did not want to let the people come out, but Hatchawa insisted on it. Puana made a thin rope and hook and dropped it into the hole. The people came out, just as many men as women. Finally a pregnant woman tried to come out and she broke the thin rope in getting out. That is the reason there are few people.

The world was dark and cold. There was no fire. Puana had made the earth and everything on it, and Itciai had created the water. Hatchawa took a live jagupa (fish) and threw it into the fire which was kept burning in the center of Kuma-land, a high circular pasture. The little fish struggled and knocked coals all about, and the people ran away in all directions with the coals. One part of these people were the Yaruros. Then Kuma wanted to give the horse to them, but the Pumeh (Yaruros) were afraid to mount it.

Of every plant in Kuma land there exists (or existed) a gigantic type, so big that an ax can't cut it. Of every animal there exists a gigantic representative.

3. India Rosa is the same as big Kuma. This Kuma lives in her city in the east. She is either the wife or sister of the sun. She is the younger sister of the other Kuma. She taught the women to make pottery and weave basketry in the same way as Puana taught the men. Itciai and the other Kuma look after everything.

4. At first there was nothing. The snake, who came first, created the world and everything in it, including the water courses, but did not create the water itself. The jaguar, the brother of the snake, created the water. The people of India Rosa were the first to people the land. After them, the other people were created. India Rosa came from the east. The Guahibos were created last. That is the reason that they live in the bush.

   Horses and cattle were given first to the Yaruros. However, they were so large that the Yaruros were afraid to mount them. The "Racionales" were not afraid, and so they were given the horse.

   The sun travels in a boat from the east. It goes to a town at night. The stars are his children and they go out from the town at night. The moon, who is a sister of the sun, also travels in a boat.

5. A woman who came from the east went to live with the sun at his village in the west. She taught women how to do everything which women do. The sun taught the men. The sun and India Rosa are married, and probably were the first people from whom everyone has sprung. But the sun and India Rosa came out of the ground. They had children. Everything was dark at that time. The children dispersed in all directions. They became the different peoples of the world. Then everything was covered with water. Horses were given to the people but they were afraid and would not ride them. But a white man sick with smallpox rode the horse, and then the horse was given to his people. He asked the Yaruros to kill him and they did. Then his people killed the Yaruros.

6. India Rosa came first. She gave birth to a son and a daughter. The son impregnated his sister, who gave birth to all

humanity. India Rosa went west, the daughter went east. The son is the sun. The moon is the daughter. The snake came afterwards, and the jaguar created the water.

7. Kuma was first. God appeared. Had two children, brother and sister, and they married. There were no human beings at that time. One day Kuma said, "Let us have some people." So God went out to see about it. He found a man in a hole. He went back to Kuma, consulted with her, and went back to the man with a hook and a rope. A pregnant woman wanted to be the first to come out of the hole, but she was left to the last. Many people were brought out. The last to be brought out was the pregnant woman, and then the rope broke. The world was dark and cold. So God made a fire. A fish appeared and scattered it, so that each person could take a little of the fire. That is why all people have fire today. The people married among themselves. One of the woman descendants of India Rosa married a man of the new race and from them sprang the Yaruros. This was welcomed because the father of the girl said, "Here, a son-in-law will take care of me now!" Then the Yaruros lived. The shaman had a nephew and a son. The nephew fell in love with his own sister and married—he was changed into a jaguar and she into a snake (?). If it had not been for this there would not have been any snakes and jaguars. Human beings should not marry their own sisters. It was ordered by Kuma. Animals are different.

Then one man found a tree with all the fruits on it. He did not tell the others. A white man appeared on horseback. Said he would come back in eight days. He came back in a boat. Scattered seeds everywhere. Thus he changed the country. Before it was all open savanna, but now forests and agricultural products grew.

India Rosa taught the women. God taught the men. God wanted to give the horse to the Yaruros, but they were afraid to mount, so he gave it to the Racionales instead.

Reprinted from Vincenzo Petrullo, "The Yaruros of the Capanaparo River, Venezuela," *U.S. Bureau of American Ethnology Bulletin 123 (Anthropological Papers, Number II),* Washington, DC: Government Printing Office, 1939, 238–241.

Source: Sproul, 305–308.

Yggdrasil

Yggdrasil is the most famous version of the **world tree,** also called the axis tree or axle tree. It is at the center of the Norse (Icelandic, Viking) creation (see also **Icelandic Creation**), and like all world trees is the axis of the various worlds of creation—the heavens, the earth, and the underworld. It joins the sky to the earth, the light to the dark, and thought to matter. It is the place of creation between the separated world parents (see also **Separation of Heaven and Earth; World Parent Creation**). Yggdrasil is an ash tree; its origins are timeless and it will survive Ragnarok, the end of the world. Snorri Snurluson tells us "its branches spread out over the whole world and reach up over heaven" (see also **Prose Edda**). The god Odin was hanged on the tree as Jesus was crucified on the Christian world tree, the cross, and as the Phrygian man-god Attis was sacrificed on the sacred pine. Yggdrasil nourishes and guards creation.
Source: Crossley-Holland, xxiii–xxiv.

Yokut Creation

Each of the branches of the Yokut tribe of central California has a version of the creation myth. These myths are earth-diver myths and most tell aspects of the **Coyote** cycle (see also **Earth-Diver Creation; Trickster**).

The Truhohi Yokut myth says that in the beginning the only land was a mountain in the south rising above the expanse of water that covered the earth. Eagle was the chief then. The people came to him asking for earth on which to live. Eagle was at a loss as to what to do, but Coyote was not; he said Magpie would know what to do. When asked, Magpie said that earth could be obtained from "right below us."

The ducks all died trying to get mud from the depths. Only Mudhen was left, and he dove and was gone for a long time before returning dead. In his nails, beak, ears, and nose, however, were bits of mud. The people made land out of chiyu seeds and this mud.

Eagle sent Wolf to make mountain ranges, and he did. Coyote disobeyed Eagle, however, and walked on them before they were dry, and that is why the mountains—especially the Sierra Nevada—are so jagged. Prairie Falcon and Raven were also sent to make mountains, and they argued about whose were higher. When everything had dried, Eagle and Coyote sent the people, who were still animals, to the different places, and soon they turned into human people.

Then only Coyote and Eagle were left in the original place. Eagle decided to go up to the heavens, and Coyote said he would do the same. Eagle ordered Coyote to stay behind to watch over things, but Coyote managed to go with Eagle anyway, even though he had no wings.

The Gashowu Yokuts claim that Prairie Falcon and Raven were the primary creators—they made the earth when there was only water everywhere. It was the duck, K'uik'ui, who managed to get sand from under the primordial waters. Prairie Falcon mixed tobacco with the bits of sand. He gave half of the mixture to Raven, and then they went in opposite directions, sprinkling the creative sand in the water to make land. Raven made better mountains, and later Prairie Falcon altered them to suit himself. It was he, after all, who had first thought of creation.

Among the Wukchamni Yokuts, Eagle and Coyote are the principle figures, as in the Truhohi myth. **Turtle** comes into this story also. It was he the creators sent to dive. He came back with a bit of sand in his nails, and out of this, Eagle and Coyote made our earth. They also made six men and six women (see also **Creation from Clay**). These they sent out as couples to populate the earth.

Later Eagle sent Coyote to see what the people were doing. It seems they were eating up the earth, so Eagle sent the dove to find something better to eat. All of the world's agriculture began with a tiny grain of meal the dove found and Eagle and Coyote planted.

Presented here is the Yauelmani Yokut myth as retold by A. L. Kroeber.

> At first there was water everywhere. A piece of wood (wicket, stick, wood, tree) grew up out of the water to the sky. On the tree there was a nest. Those who were inside did not see any earth. There was only water to be seen. The eagle was the chief of them. With him were the wolf, Coyote, the panther, the prairie falcon, the hawk called *po'yon,* and the condor. The eagle wanted to make the earth. He thought, 'We will have to have land.' Then he called *k'uik'ui,* a small duck. He said to it: 'Dive down and bring up earth.' The duck dived, but did not reach the bottom. It died. The eagle called another kind of duck. He told it to dive. This duck went far down. It finally reached the bottom. Just as it touched the mud there it died. Then it came up again. Then the eagle and the other six saw a little dirt under its finger nail. When the eagle saw this he took the dirt from its nail. He mixed it with *telis* and *pele* seeds and ground them up. He put water with the mixture and made dough. This was the morning.

Then he set it in the water and it swelled and spread everywhere, going out from the middle. (These seeds when ground and mixed with water swell.) In the evening the eagle told his companions: 'Take some earth.' They went down and took a little earth up in the tree with them. Early in the morning, when the morning star came, the eagle said to the wolf: 'Shout.' The wolf shouted and the earth disappeared, and all was water again. The eagle said: 'We will make it again,' for it was for this purpose that they had taken some earth with them into the nest. Then they took *telis* and *pele* seeds again, and ground them with the earth, and put the mixture into the water, and it swelled out again. Then early next morning, when the morning star appeared, the eagle told the wolf again: 'Shout!' and he shouted three times. The earth was shaken by the earthquake, but it stood. Then Coyote said: 'I must shout too.' He shouted and the earth shook a very little. Now it was good. Then they came out of the tree on the ground. Close to where this tree stood there was a lake. The eagle said: 'We will live here.' Then they had a house there and lived there.

Reprinted from A. L. Kroeber, *Indian Myths of South Central California,* University of California Publications, American Archaeology and Ethnology, IV, no. 4 (1906–1907): 229–231.

Source: Long, 208–214.

Yoruba Creation

The Yoruba live in what is now Nigeria. Their creator is Olurun or Olodumare. Sometimes he is assisted by the lesser god, Obatala.

It is said that in the beginning water was everywhere and Olurun, the supreme being, sent Obatala, or Orishanla, down to create some land from the chaos (see also **Creation from Chaos**). He went down on a chain and took a shell with some earth, some iron, and a rooster (some say a pigeon hen) in it. He put the iron in the waters, the earth on top of the iron, and the cock on top of the earth. The cock's scratching spread the land about, and when it was ready, some other lesser gods came down to live there with Obatala. Chameleon came first to see that it was dry enough. When it was, Olurun named earth Ife, meaning wide.

Obatala created humans out of earth (see also **Creation from Clay**) and got Olurun to blow life into them. Then one day he got drunk and by mistake started making cripples, who are now sacred to Obatala. Some say that Obatala was jealous of Olurun and wanted to give humans life by himself, but Olurun put him to sleep while he was working so he saw nothing. They say, however, it is Obatala who shapes babies in their mothers' wombs.
Sources: Beier, 47; Hamilton, 73–77.

Yuchi-Creek Creation

The Yuchi Indians were deported to Oklahoma from the South along with the "Five Civilized Tribes"—the Cherokee, Chichasaw, Choctaw, Creek, and Seminole. This myth was passed on from the Yuchis to the Creeks and now belongs to both tribes. An earth-diver type story, it has remnants of a matrilineal point of view, since the sun is female and the serpent—always associated with the Great Goddess in ancient times—plays a significant role (see also **Earth-Diver Creation**).

In the beginning water was everywhere. Crawfish dove down to the bottom to find mud. He stirred some up and took it away, but the mud people down below were cross. "Who is bothering our mud?" they cried. Crawfish, however, moved so fast and stirred up so much mud that the mud people could never catch him. He brought up more and more land.

By flapping his great wings over the land, Buzzard stirred up mountains, made valleys, and dried things out.

Next Yohah, the star, gave light to the land. It was not light enough, however, so Moon added his light. Finally, the Great Mother (the sun), gave her light. She moved across the sky each day, and one day a drop of her blood fell to the earth and gave birth to the first Yuchis (Uchees).

The Yuchis were bothered by a great serpent. They cut off its head but it grew back. When they cut off its head again, they put it at the top of a tree; it killed the tree, however, and found its body again. Only when they put the severed head on the cedar tree did the monster die. This ceremony of the head and the cedar tree is the basis of Yuchi medicine.

The people acquired fire and language, and they lived happily together. The Yuchi are the Tsohaya, the People of the Sun, and they place a picture of the sun over each of their dwellings.
Sources: Leach, 88–89; Sproul, 255–257.

Yuki Creation

The Yuki Indians live in Round Valley of northern California. Their creator, Taiko-mol (Solitary Walker), is an example of **creation by *Deus Faber*,** who as a craftsman constructs the world from chaos (see also **Creation from Chaos**). In some versions of the myth, Coyote is an observer of creation.

In the beginning there was foam that wandered around on the surface of fog-covered waters. A voice came from the foam followed by Taiko-mol, who had eagle feathers on his head. The creator stood on the moving foam and sang as he created. In the darkness he made a rope, and he laid it out on the north-south axis. Then he walked along it, coiling it and leaving the created earth behind him as he went. He did this four times, and each time the water overwhelmed the new land. As he walked he wondered if there was a better way. Then he made four stone posts or *lilkae* and secured them in the ground in each of the four directions. He attached lines to these and stretched them out across the world as a plan. Finally he spoke the Word (see also **Creation by Word**) and the earth was born. Then the creator secured the new world from the waters by lining it with whale hide. He shook the earth to see that it was indeed secure—this was the first earthquake. Earthquakes since then are Taiko-mol retesting his work.

Source: Weigle, 180–181.

Yuma Creation

The Yuma Indians of Arizona tell an emergence/earth-diver creation story that stresses the natural and original struggle between good and evil in creation (see also **Creation by Emergence; Earth-Diver Creation**).

At first there was only water and emptiness. Then mist from the waters became sky. Then the Creator, who lived without form deep in the maternal waters, was born of those waters as the twins, Kokomaht, the good one, and Bakotahl, the evil one. As he came up through the waters, Kokomaht kept his eyes closed. Before he came out, Bakotahl cried out to his brother to ask whether he had opened or closed his eyes during the passage. The good twin knew the evil nature of his other half, so he lied and said his eyes had been open. So Bakotahl opened his eyes as he came through and was blinded, as Kokomaht had known he would be. This is why he is named Bakotahl, or Blind One.

Kokomaht set about making the four directions, taking four steps on the water in each direction and pointing and announcing the names: north, south, east, west (see also **Creation by Word**).

Now Kokomaht said he would make the earth, but Bakotahl doubted his twin's power. "Let me try first," he said. "No," said Kokomaht, and he stirred up the waters so much that they brought up land. Kokomaht sat on the land.

Bakotahl was angry at his twin, but sat down next to him. Secretly he made a little human figure out of mud (see also **Creation from Clay**), but it was imperfect, to say the least.

Kokomaht himself decided to make a new being, and he made a perfect man, who got up and walked. Then he made a perfect woman.

Bakotahl continued his imperfect work as well and told his twin that what he had made were people. Kokomaht pointed out the imperfections of his brother's work—no hands, no feet. Bakotahl was so angry that he dove back into the depths and sent up storms, which Kokomaht stomped out, but not before sickness slipped into the world.

The first man and woman made by Kokomaht were the Yuma ancestors. Kokomaht went on to make the ancestors of other tribes as well—the Dieguenos, Apaches, Pimas, and others. He made 24 pairs of humans before he finally made white people.

Kokomaht taught the Yumas how to live, especially how to have children. He made a son himself out of the void, without a woman. This boy was Komashtam'ho, and he taught the people how to make children by joining together, male and female.

Kokomaht sensed that it was too dark in the world, so he made the stars and the moon. Then he announced that he had done what he could and that his son would continue his work.

It was Frog who was jealous of Kokomaht and decided to kill him. Kokomaht knew what his people thought, however, so he knew Frog's intentions. He decided that death must be a part of creation and that he would use himself to begin it. He allowed Frog to murder him by sucking out his breath. Then he lay down to die. He called the people to gather around him; only the white man stayed away, pouting over his washed-out looks and anxious to grab whatever he could grab. To quiet the white man, Komashtam'ho made him a horse out of sticks.

In his final talk with the people, Kokomaht taught them about dying. Then he died himself.

Komashtam'ho continued with creation. First he made the sun and then wood. With the wood he made a funeral pyre for his father. He sent **Coyote** to get a spark from the sun, but as soon as Coyote was gone, the good twin made fire with sticks and lit the pyre (see also **Trickster**). As the body was burning, Coyote stole its heart. For this theft, he was condemned to be a wild man and a thief.

Komashtam'ho explained to the people about death and the afterlife. In the world after, the people would be strong and happy and would be with those who had gone before them.

Komashtam'ho chose the man Marhokuvek to assist him in his continuing creation. Marhokuvek told the people to cut their hair in mourning for the creator, but when he saw how silly the animals looked in their shaved form—animals had once looked the same as the previously hairy humans—he gave them back their hair, which they still have now.

Komashtam'ho sent a flood to rid the world of some of the wilder animals, but good animals and humans died too. Marhokuvek pleaded for mercy (see also **The Flood and Flood Hero**), and Komashtam'ho sent a fire to dry up the water. This is why we have deserts in Yuma country.

Then, after teaching more about death and after creating the Colorado River, mountains, and many other things, Komashtam'ho turned himself into four eagles to watch over the people.

As for Bakotahl, he is still under the earth causing trouble for the people above. Bad things come from him, as good things come from Kokomaht.
Source: Erdoes, 77–82.

Yurucare Creation

In Bolivia, the Yurucare Indians believe that the demon Aymasune brought fire down from the skies and burned up everything, including the human race. This is a fire version of the Great Flood (see also **The Flood and Flood Hero**). One man was saved from the fire storm; he hid in a cave, holding out a stick at regular intervals to see if the fire had stopped. On the third try the stick was cool, and after four more days the man came out to see the blackened world and to begin a new life.
Source: Freund, 10.

Zambian Creation

See Malozi Creation.

Zia Creation

The Zia pueblo tells an emergence creation myth (see also **Acoma Creation; Creation by Emergence**). The main deity is Tsityostinako or Prophesying Woman, who, since she is sometimes said to be a spider, is probably related to **Spider Woman.**

In the beginning Prophesying Woman lived alone with Utctsiti and Naotsiti, her two daughters. At that time there was a lot of fog and four worlds: the yellow one at the bottom, the blue one next, followed by the red one, and the white one on top. Tsityostinako and her daughters were still at the bottom.

The daughters used a magic blanket, a *manta,* to create things. They sat on the floor with the manta open in front of them and a magic cane on it, and they created. Prophesying Woman was invisibly with them giving them ideas. They sang to help the creation along and checked regularly under the manta to see what was going on there. Later, when the people were created, they made their way up through the various worlds.
Source: Weigle, 33.

Zoroastrian Creation

Zoroastrianism is a religion of Persia (see also **Persian Creation**). It was founded by the historical Zarathustra, or Zoroaster, born in the sixth century B.C.E. The religion is related to older Persian traditions, especially Mithraism, the cult of the Indo-Iranian sun god Mithra. In its emphasis on the great dualities of light and dark (good and evil), Zoroastrianism influenced Christianity. Had Xerxes and his Persian navy defeated the Greeks in the great naval battle of 480 B.C.E., Zoroastrianism, rather than Christianity, might well be the dominant religion of the Western world today.

The Avesta, written in the ancient and difficult Avestan language, is the sacred book of Zoroastrianism. Tradition has it that Zoroaster composed the text himself, but it was not completed in the form in which we have it now until some two hundred years after the prophet's death.

Zoroastrianism stresses the necessity of joining with the good forces of nature. Ahura Mazda is the high god, the omnipotent and omniscient creator. Against him stands Angra Mainyu, the force of evil. Earth is holy for the Zoroastrians, and righteousness is the duty of the believer.

The Avesta, in which we find the earliest form of the religion, sees the source of matter in an essential purity. First there was Light—alone and without end. Within the Light was the power of the Word (see also **Christian Creation**) and the power of Nature. Our world came about when the creator joined together the spirits of the Word and Nature (see also **Creation by Word**).

In the beginning the new creation was chaos, but out of the union of Word and Nature came true form as the universe (see also **Creation from Chaos**).

In about 350 C.E., an official analysis of the Avesta was compiled. What follows is a description of creation from that work.

> Concerning the original coming to be of the material creation, from the Exegesis of the Good Religion.
>
> This is what is revealed concerning the instrument which the Creator fashioned from the Endless Light and in which he caused creation to be contained. Its Avestan name is the Endless Form and it is twofold. On the one hand it contains the "ideal" creation; on the other the material creation. In the "ideal" creation the Spirit of the Power of the Word was contained; and in the material creation the Spirit of the Power of Nature was contained, and it settled (in it). The instrument which contains the ideal creation was made perfect and the spiritual (ideal) gods of the Word were separated from it, each for its own function, to perform those activities which were necessary for the creation that was within the instrument. And within the

Zarathustra, a Persian born in the sixth century B.C., *was known to Greeks as Zorastar. He compiled the Avesta, which included a creation story in which Zervan, center, created twins, to his left: Ahura Mazda, who stood for good, and the evil Angra Mainyu.*

instrument which contains the material creation the marvellous Spirit of the Power of Nature was united to the kingdom of the Spirit of the Power of the Word through the will of the Creator.

In greater detail this is what is revealed concerning the nature of the material world. First was the stage the Avestan name of which is the mass and which is called the . . . mass or conglomeration in the language of the world. From the stage which is called the mass and conglomeration (proceeded) the stage the Avestan name of which is conception and also hollowing and which is called the new phenomenon(?) and also . . . in the language of men. It is (as) an ailment(??) in the mass, and it was inserted into it. From the stage which is called conception, hollowing, the new phenomenon(?) and . . . (proceeded) the stage the Avestan name of which is formation and which is called expansion in the language of men. It is as an ailment(??) in the (former) stage and it was inserted into that stage. From the stage the name of which is formation and expansion (proceeded) the first body united with the Spirit of the Power of the Word; and in the language of men it is called the firmament *(spihr).* In it, like embryos, are the luminaries, the Sun, Moon, and stars, all of the same origin. They control all creation under them and are themselves the highest natural phenomena. From the Wheel proceeded becoming, the hot and the moist of which air is composed: these are connected with the spiritual Word and share in its power: (and from it also proceeded) the elements which are the seed of seeds of material creations and which are the movement of becoming. The elements are called the form of the becoming of the primary qualities. From the movement of becoming (proceeded) the settling of becoming, living things, which include material cattle and men, for they are the form and shape of matter.

In the fifth century B.C.E., the concept of Ahura Mazda as the creator, the combined power of the two spirits of the original creation, developed.

Ahura Mazda was sometimes called Ohrmazd, and his opposite, Angra Mainyu, was Ahriman. It was written that Ohrmazd created Finite Time (Zurvan) from Infinite Time and placed the stars in it to measure its passing. Ahriman was weak during the first 3,000 years of the 12,000 years of Finite Time, but after the next 3,000 years, during which the world was made and Adam and Eve were created, Ahriman's powers polluted creation and Ahriman had to be contained in Hell.
Source: Sproul, 135–140.

Zulu Creation

Unkulunkulu, the Ancient One, is the Zulu creator. Nobody knows where he is now; he came originally—that is, he "broke off"—from some reeds, which play the role originally played by the cosmic egg (see also **Creation from a Cosmic Egg**). Some say that he was the reeds, because the word for them, *uthlanga,* means source. It was he who broke off the people from the reeds and then the cattle and other peoples. He also broke off medicine men and dreams. He was really the first man and the progenitor of other men.

Unkulunkulu created everything that is—mountains, cattle, streams, snakes. He taught the Zulu how to hunt, how to make fire with sticks, and how to eat corn. He named the animals for them.

The people say that Unkulunkulu is in everything; Unkulunkulu is the corn, the tree, the water. Some say that a woman followed him out of the original reeds, then a cow and a bull, then the other pairs of animals. Whatever the story, Unkulunkulu was the first man and there was nothing before him; yet he broke off from the source.
Source: Leach, 148.

Zuni Creation

The Zuni pueblo in New Mexico is the home of a well-preserved and viable culture. Zuni ceremonies and myths reflect that viability in their intensity and complexity. The creation myth is no exception. Like the creation stories of the other pueblos, it is an emergence myth (see also **Creation by Emergence**). It is also an example of *ex nihilo* creation (see also **Creation from Nothing**) and of **creation by thought.**

Awonawilona is the creator of all that is. He existed before anything else in the great dark emptiness of the beginnings. He conceived himself by thought; as the container of all things, he created himself as himself and as the sun that brought people light, warmth, and water.

Out of himself Awonawilona made the seed with which he impregnated the primeval waters with Awitelin Tsita (Earth Mother) and Apoyan Ta'chu (Sky Father). Sky Father and Earth Mother came together and engendered the creatures of our world (see also **World Parent Creation**). Then Earth Mother cast off Sky Father (see also **Separation of Heaven and Earth**) and sank in comfort part way into the waters.

Sky Father and Earth Mother can take many forms; like thoughts, they can transform themselves at will, the way dancers can in the ceremonies. So it is that they could speak to each other as humans and see things in human terms.

Earth Mother held up a great bowl of water and told Sky Father the bowl was herself and that along its rim the people would live. She said she was many bowls for many peoples, many countries. The people, she said, would take nourishment from her as from the water of the bowl. Earth Mother spat into the bowl and stirred it, causing land to build up on the edges. Sky Father sent his cold breath down, which made clouds and mist on the land. The children of earth would seek shelter in her lap, said Earth Mother. The Earth Mother is warm, say the people, and the Sky Father is cold. So it is with women and men.

Sky Father, too, gives to humans. He passed his hand over the bowl and up sprang the fruits of the earth.

The Zunis also talk of the beginnings from a less mystical and less cosmic point of view. They tell how the people actually came to be in Zuni.

The Fourth World was dark, they say, and crowded. The people constantly got in each other's way. In the world there was the sun and the earth. The Father looked down on the beautiful Mother and pitied the people crowded in the darkness below her. He sent his rays down to earth in such a way as to encourage his two sons, Elder Brother and Younger Brother, to go in search of the people. "Let's go find the people, so they can come up here and see our father," said Younger Brother.

The brothers went southwest until they came to the entrance to the below. It was dim in the First World, dark in the Second World, very dark in the Third World, and pitch black in the Fourth World, where the people were. The people sensed that strangers were among them; they touched the children of the Sun in the darkness.

"Come with us and we will take you to the Sun," the brothers said, "but we must show you how."

So Younger Brother went north and planted pine seeds, turned around while a pine tree grew, and took a branch to the people. Then he did the same with the spruce in the west, the silver spruce in the south, and the aspen in the east.

The sons of the Sun built a prayer stick ladder of the pine tree from the north for the people to use on their climb to the third world. They stayed there for some time—some say four days, some say four years. The sons then made a prayer stick ladder out of the western spruce and the people climbed to the second world. The dim light there almost blinded them. They stayed there for a while; then the sons made a prayer stick ladder from the silver spruce of the south and the people climbed to the first world, where they had to cover their eyes in the dawnlike light. When they could see, the people were horrified by the way they looked; they were dirty and slimy and they had tails, no mouths, and webbed feet. They remained in the first world for four days (or four years).

Finally the sons of the Sun made a prayer stick ladder out of the aspen tree of the east, and, accompanied by thunder, the people climbed out into our world. The brightness brought tears of pain to their eyes, but Younger Brother forced them to look directly at the Father Sun. The tears of pain flowed, and from these tears came flowers.

"Now you are in the world," said the sons.

The people rested at the emergence spot for four days (or four years) before moving on to Awico, where the sons taught the people how to grow food. The people liked the way the corn smelled, but they had no mouths, so the brothers cut mouths into them while they slept. Later they cut the webs between the people's fingers and outlets in their bottoms. Now the people could handle the food, eat it, and give off waste.

One night the brothers removed the tails and horns from the people. Some woke up and asked to keep their tails; these became monkeys. Most of the people were pleased with the way they looked; they looked like us.

Sources: Leach, 65–71; Sproul, 284–286.

Bibliography

Beckwith, Martha Warren. *Mandan-Hitatsu Myths and Ceremonies. Memoirs of the American Folklore Society.* American Folklore Society, 1937.

Beier, Ulli. *The Origin of Life and Death: African Creation Myths.* London: Heinemann, 1966.

Berndt, R. M. "Some Aspects of Jaralde Culture, South Australia," *Oceania* XI, 2 (1940): 164–185.

Bierhorst, John. *The Mythology of North America.* New York: Morrow, 1985.

Boer, Charles, trans. *The Homeric Hymns.* Chicago: Swallow, 1970.

Brandon, S. G. F. *Creation Legends of the Ancient Near East.* London: Hodder and Stoughton, 1963.

Brown, Charles E. *Wigwam Tales.* Madison, WI.: University of Wisconsin Press, 1930.

Clark, R. T. Rundle. *Myth and Symbol in Ancient Egypt.* London: Thames and Hudson, 1959.

Colum, Padraic. *Myths of the World [Orpheus].* New York: Grosset and Dunlap, [1930] 1972.

Crossley-Holland, Kevin. *The Norse Myths.* New York: Pantheon, 1980.

Curtin, Jeremiah. *Creation Myths of Primitive America: In Relation to the Religious History and Mental Development of Mankind.* Boston: Little, Brown, 1898.

Cushing, Frank H. *Outlines of Zuni Creation Myths.* Bureau of American Ethnology Annual Report, No. 13. Washington, DC: United States Government Printing Office, 1896.

Dalley, Stephanie, ed. *Myths from Mesopotamia: Creation, the Flood, Gilgamesh, and Others.* New York: Oxford University Press, 1989.

Dange, S. S., ed. *Myths of Creation.* Bombay: South Asia Press, 1987.

Davidson, H. R. Ellis. *Gods and Myths of Northern Europe.* Middlesex: Penguin, 1964.

Del Re, Arundel. *Creation Myths of the Formosan Natives.* Tokyo, Japan: Hokuseido, 1975.

Dundes, Alan, ed. *The Flood Myth.* Berkeley: University of California Press, 1988.

Eliade, Mircea, ed. *The Encyclopedia of Religion.* 16 vols. New York: Macmillan, 1987.

———. *From Primitives to Zen: A Thematic Source Book of the History of Religions.* New York: Harper & Row, 1974 [Part I of as *Gods, Goddesses, and Myths of Creation.* New York: Harper & Row, 1974].

———. *Myth and Reality.* New York, 1963.

———. *The Myth of the Eternal Return* (1954). Princeton, NJ: Princeton University Press, [1954] 1971.

———. *Patterns in Comparative Religion.* New York: Steed & Ward, 1958.

———. *The Sacred and the Profane: The Nature of Religion....* New York: Harcourt, Brace and World, 1959.

Embree, Ainslie T. *Sources of Indian Tradition.* Vol I. New York: Columbia University Press,1988.

Emory, Kenneth, trans. "Cosmic Chant from the Takaroa Atoll of the Tuamotus," *Journal of the Polynesian Society* 47, 5 (1920): 14.

Erdoes, Richard, and Alfonso Ortiz. *American Indian Myths and Legends.* New York: Pantheon, 1984.

Fallon, Francis T. *The Enthronement of Sabaoth: Jewish Elements in Gnostic Creation Myths.* Leiden, Netherlands: Brill, 1978.

Farmer, Penelope, ed. *Beginnings: Creation Myths of the World.* New York: Atheneum, 1979.

von Franz, Marie Louise. *Patterns of Creativity Mirrored in Creation Myths.* Zurich, Switzerland: Spring Publications, 1972.

Freund, Philip. *Myths of Creation.* New York: Washington Square Press, 1965.

Frye, Northrop. *Creation and Recreation.* Toronto: University of Toronto Press, 1980.

Gowan, Donald E. *Genesis 1–11: From Eden to Babel.* Grand Rapids, MI: Erdmans, 1988.

Graves, Robert. *The Greek Myths.* 2 vols. Baltimore, MD: Penguin, 1955.

Hamilton, Virginia. *In the Beginning: Creation Stories from around the World.* New York: Harcourt Brace Jovanovich, 1988.

Hesiod. *Hesiod and Theogony.* Harmondsworth, England: Penguin, 1973.

The Holy Bible. King James version. New York: Harper & Brothers, n.d.

Kilpatrick, George W. *The Bengal Cosmogony: Some Creation Myths in the Literature of the Dharma Cult.*

Kramer, Samuel Noah, ed. *Mythologies of the Ancient World.* Garden City, NY: Doubleday, 1961.

———. *Sumerian Mythology.* New York: Harper & Row, 1961.

Kramer, Samuel Noah, and John Maier. *Myths of Enki, The Crafty God.* New York: Oxford, 1989.

Kroeber, A. L. "Indian Myths of South Central California." University of California Publications. *American Archaeology and Ethnology* 4, 4 (1906–1907): 229–231.

Leach, Maria. *The Beginning: Creation Myths around the World.* New York: Thomas Y. Crowell, 1956.

Leeming, David A. *The World of Myth.* New York: Oxford, 1990.

Locke, Raymond Friday. *The Book of the Navajo.* Los Angeles: Mankind, 1989.

Long, Charles H. *Alpha: The Myths of Creation.* New York: George Braziller, 1963.

Lönnrot, comp. and Francis Magoun, Jr., trans. *The Kalevala.* Cambridge, MA: Harvard University Press, 1963.

Lovelock, James. "Gaia: A Model for Planetary and Cellular Dynmaics," in *Gaia: A Way of Knowing,* ed. William Erwin Thompson. New York: Lindisfarne, 1987.

Maclagan, David. *Creation Myths: Man's Introduction the World.* London: Thames & Hudson, 1977.

Marriott, Alice, and Carol K. Rachlin, eds. *American Indian Mythology.* New York: Mentor, 1968.

———, eds. *Plains Indian Mythology.* New York: Mentor, 1975.

Mascaro, Juan, trans. *The Upanishads.* Baltimore, MD: Penguin, 1965.

Mason, Herbert, trans. *Gilgamesh: A Verse Narrative.* New York: Mentor, 1970.

Middleton, John, ed. *Myth and Cosmos: Readings in Mythology and Symbolism.* Garden City, NJ: Natural History Press, 1967.

Morford, Mark P. O., and Robert J. Lenardon. *Classical Mythology.* 3d ed. New York: Longman, 1985.

Mullett, G. M. *Spider Woman Stories.* Tucson, AZ: University of Arizona Press, 1982.

O'Brien, Joan, and Wilfred Major. *In the Beginning: Creation Myths from Ancient Mesopotamia, Israel, and Greece.* Chico, CA: Scholars Press, 1982.

O'Flaherty, Wendy Doniger, ed. *Hindu Myths.* New York: Penguin, 1975.

Olcott, William T. *Myths of the Sun.* New York: Capricorn Books, 1914.

Oleyar, Rita. *Myths of Creation and Fall.* New York: Harper & Row, 1975.

Ovid, *Metamorphoses.*

Petrullo, Vincenzo. "The Yaruros of the Capanaparo River, Venezuela." *U.S. Bureau of American Ethnology Bulletin* 123 (Anthropological Papers, no. II). Washington, DC: U.S. Government Printing Office, 1939.

Pritchard, James B., ed. *Ancient Near Eastern Texts Relating to the Old Testament.* Princeton, NJ: Princeton University Press, 1950.

Rothenberg, Jerome, ed. *Shaking the Pumpkin: Traditional Poetry of the Indian North Americas.* Garden City, NY: Doubleday, 1972

Sproul, Barbara C. *Primal Myths: Creation Myths around the World.* San Francisco: HarperCollins, [1979] 1991.

Sturluson, Snorri. *The Prose Edda.* Translated by Jean Young. Berkeley: University of California Press, [1954] 1973.

Swimme, Brian. *The Universe Is a Green Dragon: A Cosmic Creation Story.* Santa Fe, NM: Bear & Co., 1984.

Tedlock, Dennis, trans. *Popul Vuh.* New York: Simon & Schuster, 1985.

Thompson, Stith. *Motif-Index of Folk Literature.* Bloomington, IN: Indiana University Press, 1955–1958.

Thompson, Stith. *Tales of the North American Indians.* Bloomington, IN: Indiana University Press, [1929] 1971.

Tyler, Hamilton A. *Pueblo Gods and Myths.* Norman, OK: University of Oklahoma Press, 1964.

Tylor, Edward B. *Primitive Culture.* New York: Harper & Row, [1871] 1960.

Underhill, Ruth. *Papago Woman.* New York: Holt, Rinehart and Winston, [1936] 1979.

Van Over, Raymond. *Sun Songs: Creation Myths from around the World.* New York: Dutton, 1980.

Walker, Barbara. *The Woman's Encyclopedia of Myths and Secrets.* San Francisco: Harper & Row, 1983.

Weigle, Marta. *Creation and Procreation: Feminist Reflections on Mythologies of Cosmogony and Parturition.* Philadelphia: University of Pennsylvania Press, 1989.

Williamson, Ray A. *Living the Sky: The Cosmos of the American Indian.* Norman, OK: University of Oklahoma Press, 1984.

Wood, Nancy. *When Buffalo Free the Mountains: The Survival of America's Ute Indians.* New York: Doubleday, 1980.

Ywahoo, Dhyani. *Voices of Our Ancestors: Cherokee Teachings from the Wisdom Fire.* Boston: Shambhala, 1987.

Zaehner, C. *Zurvan: A Zoroustrian Dilemma.* Oxford: Clarendon Press, 1955.

Zolbrod, Paul G. *Dine bahane: The Navajo Creation Story.* Albuquerque, NM: University of New Mexico Press, 1984.

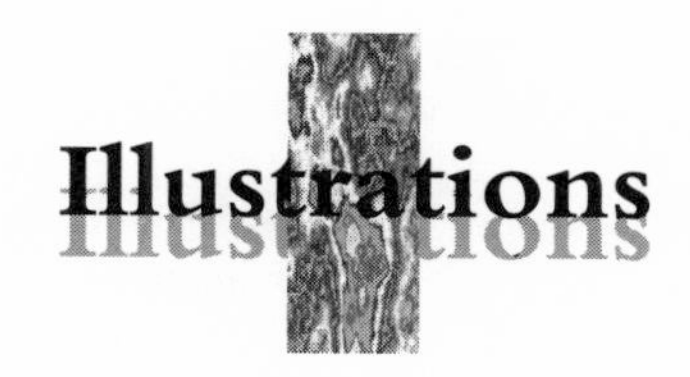

Illustrations

5 *Elohim Creating Adam.* Watercolor by William Blake. Tate Gallery, London/Art Resource S0031602, New York.

19–20 Bark paintings by Wurungulngul. Berndt Museum of Anthropology WU/P 1988 and WU/P 1989, University of Western Australia, Nedlands, Western Australia.

21 Codex Borgia, folio 32 (facsimile). Department of Manuscripts, British Library, London.

24 Babylonian Creation Tablet I, Western Asiatic Collection 93014, British Museum, London.

32 PML 12 ChL ff1 Biblia Latina, vol. I. Mainz: Gutenberg & Fust (ca. 1455) folio 5, M.500, f.4v. The Pierpont Morgan Library, New York.

36 Birth of Buddha relief (second century). Indian Museum, Calcutta, India. Giraudon/Art Resource S0031619, New York.

45 *Legend of Creation—1.* Painting by Jimalee Burton. Courtesy Cherokee National Museum, Tahlequah, Oklahoma. Photograph by Sammy Still.

48 Department of Oriental Antiquities 1962. 12-31-02 (20), British Museum, London.

51 Codex Sinaiticus, folio 260. Department of Manuscripts Add. 43725, British Library, London.

57 Codex Vindobonensis 2554, folio 1v. Bild-Archiv der Osterreichische Nationalbibliothek, Vienna, Austria.

70 *Djanggawul Creation Story.* Bark painting from Yirrkala, Arnhem Land, Northern Territory, Australia (1959). Art Gallery of New South Wales P68.1959, Sydney, Australia.

72 Musee de l'Homme, Paris. Giraudon/Art Resource S0044557, New York.

85 Akkadian cylinder seal 202. The Pierpont Morgan Library, New York.

96 *Manafi al-Hayawan* M.500, folio 4v. The Pierpont Morgan Library, New York.

102 Egyptian papyrus, 21st Dynasty. Egyptian Museum, Cairo, Egypt. Giraudon/Art Resource S0017709, New York.

105 Cretan Mother Goddess. Archaeological Museum, Heraklion, Greece. Scala/Art Resource S0031613, New York.

114 Biblia Italiana, Venice, Italy (1494). Rosenwald Collection, Rare Book and Special Collections Division, Library of Congress, Washington, D.C.

134 *The Giant Ymir and the Cow Audhumla.* Painting by N. A. Abilgaard (ca. 1777). Statens Museum for Kunst, Copenhagen, Denmark.

149 *Izanami and Izanagi Creating the Japanese Islands* by Kobayashi Eitaku. Bigelow Collection, Museum of Fine Arts 11.7972, Boston.

169 From Book I of *Walam Olum or Red Score: The Migration Legend of the Lenni Lenape or Delaware Indians.* Unpublished manuscript by Constantine S. Rafinesque (1833). Department of Special Collections MS Brinton Br 497.11DW154, Van Pelt-Dietrich Library, University of Pennsylvania, Philadelphia.

182 Hamburg Folk Art Museum, Hamburg, Germany.

187 Codex Zouche-Nuttall (facsimile edition), folio 36. Department of Manuscripts, British Library, London.

194 Koran. Oriental and India Office Collections Or 1009, folios 115v-116r, British Library, London.

199 *The Emergence.* The Wheelwright Museum of the American Indian PS21-14, Santa Fe, New Mexico.

203 Tangaroa. Ethnography Department LMS 19, British Museum, London.

227 Denver Art Museum 1979.3.

259 Louvre, Paris/Art Resource S0042803, New York.

267 Painting by Felipe Davalos. National Geographic Society.

279 American Museum of Natural History 315731, New York.

305 Silver plaque, Luristan, Iran. Cincinnati Art Museum 1957.29.

Index